THE ETHNOGRAPHER'S WAY

the
ethnographer's
WAY

a handbook *for* multidimensional research design

Kristin Peterson & Valerie Olson

DUKE UNIVERSITY PRESS   *Durham and London*   2024

© 2024 DUKE UNIVERSITY PRESS
All rights reserved
Printed in the United States of America on acid-free paper ∞
Project Editor: Bird Williams
Designed by A. Mattson Gallagher
Typeset in Cronos Pro and Literata TT by
Westchester Publishing Services
Printed and bound by CPI Group (UK) Ltd, Croydon, CR0 4YY

Library of Congress Cataloging-in-Publication Data
Names: Peterson, Kristin, [date] author. | Olson, Valerie, [date]
author.
Title: The ethnographer's way : a handbook for multidimensional
research design / Kristin Peterson, Valerie Olson.
Description: Durham : Duke University Press, 2024. | Includes
bibliographical references and index.
Identifiers: LCCN 2023026792 (print)
LCCN 2023026793 (ebook)
ISBN 9781478030157 (paperback)
ISBN 9781478025900 (hardcover)
ISBN 9781478059141 (ebook)
Subjects: LCSH: Anthropology—Research—Methodology. |
Ethnology—Research—Methodology. | Research—Methodology. |
BISAC: SOCIAL SCIENCE / Anthropology / Cultural & Social |
SOCIAL SCIENCE / Sociology / General
Classification: LCCN GN42 .P484 2024 (print)
LCCN GN42 (ebook)
DDC 301.072/1—dc23/eng/20231025
LC record available at https://lccn.loc.gov/2023026792
LC ebook record available at https://lccn.loc.gov/2023026793

CONTENTS

TABLES, EXAMPLES, FIGURES, AND FORMULAS

Tables

Examples

Figures

Formulas

Why and How to Use This Handbook

How can researchers turn inspired hunches into full-fledged projects? Since 2014, we've been teaching ethnographic research design from its nascent stages, namely imagining, conceptualizing, and clarifying project intentions. Typically, this early phase of conceptualization is a solitary and free-form process. It sets the stage for the project's innovative potential. But it is also the phase in which grave flaws can be made, such as an insufficient literature review and failure to create congruence between research questions and field-work plans. Research design often takes a backseat to the more prominently formalized processes of methods selection, proposal writing, fieldwork, data collection, and publication.

In contrast to books that deal with those more visible processes, this handbook pays attention to research design's incipient conceptualization processes and uses them to complete research question development and data-collection planning. Because we focus on conceptualization, this is not a methods handbook, but it expects designers to have been exposed to ethnographic methods and analysis training. After you've completed all the handbook's modules, you will have a framework for writing grants and other proposals, collecting data, approaching qualitative analysis, and offering important theory-making avenues in your written, audio, or visual products. But our intent goes further. We want to showcase research design's possibilities as a communally shared and iterative process *from its nascent beginnings*. This handbook, then, offers more than a way to design research—it invites ways to change research life.

We refer to our process as *research design* while keeping in mind that *design* is the focus of critiques as well as liberatory activism. Scholars in several disciplines show how the term legitimates the visions, practices, and hierarchical structures of elite engineering and architecture enclaves.[1] Their critiques often focus on the rise of *design thinking,* a planning and self-fashioning practice based on tech-market logics.[2] At the same time, ethnographers cite rising movements to reclaim and reshape design as a powerful, collective so-cial transformation activity.[3] Our experience has shown that students at all levels are empowered by the sense of purposeful making and signifying that design invites. We are inviting the reconceptualization of both research and design, based on practices of community care and open-ended ways of find-ing things out and writing about them.

The way that we offer here is one among many ways to imagine and design frameworks of ethnographic inquiry. What is "new" about it is how we tested, assembled, and refined bits and pieces of different tools and design ap-proaches into a step-by-step, iterable process. Like the design process we advo-cate, the handbook itself emerged iteratively through years of responding to student and colleague feedback on what worked and what didn't. Moreover, we attended to the experiences of recently returned fieldworkers and graduated students who were turning their projects into books, becoming instructors, and finding themselves mentoring others in research design. We also used its techniques to guide well-established researchers to design and coordinate col-lective projects made up of multiple teams and research objectives. Therefore, this book is the product of ongoing relationships with beginning and advanced designers and research collectives. We were motivated by our shared concerns with navigating the highs and lows of institutionalized research processes and managing careers in a variety of settings, academic and beyond.

Foundational to this handbook is a design mode that we call **multidi-mensional:** an *iterative approach to assembling diverse research concepts and intentions within a congruent framework of inquiry*. Unlike site, scale, or per-spective, the term *dimension* does not solely refer to situated differences—such as a place, size, or view—within a kind or category. Dimensions are broadly and differentially aspectual: they can be material or perceptual, spatial or temporal, quantifiable or immeasurable, tangible or intangible, concrete or speculative. We use *multidimensional* to signal the fully lively form that a proj-ect can take when researchers work on all its aspects and angles in a creative way. For example, one researcher we worked with, Tariq Rahman, created an ethnographic project centered on the seemingly *concrete* concept of land-as-property, which takes measurable form within the categorical dimensions of economics and legality. But researchers can also attend to people's emerging experiences of land as a *shifting* and less concretized dimension of speculative technical and spiritual space-making. The result of putting these seemingly disparate *dimensions of land* together is a multidimensional project that fo-cuses on important nonconcrete and more-than-territorial aspects of land-as-property. This is one example of putting multiple conceptual dimensions together in a way that creates intellectually compelling and socially responsive multidimensionality.

The approach we take, then, is unlike most traditionally prescribed research design procedures because it emphasizes checking out conceptual connections that can be obscured by institutionalized design ideologies. By this, we mean prescribed ideas about which kinds of people, places, processes, things, situations, contexts, and theoretical ideas designers *should* put together. To help designers do this, the handbook favors a slow try-and-refine design process. It encourages designers to work with design elements in new, nonlinear, more iterative ways. These elements include seemingly disparate concepts, theoretical perspectives, social processes, forms of data, and literatures, as well as personal experiences, imaginative intuitions, and political commitments. From a normative perspective, the project elements we coach learners to connect may sometimes seem unrelatable and incongruous, and the ways we help them make those connections may appear unconventional. But we have found that fostering this creative and often audacious try-and-see process helps designers develop innovative projects with broad intellectual significance and social impact.

Experienced researchers will recognize the spirit of multidimensional design. It is found in the intuitive analytic moves that ethnographers make in final written works in which they claim to "bring different literatures together," "connect different processes," or "juxtapose sites" in novel ways. While most anthropologists aim to design such richly multidimensional projects, the cryptic adage "you know an innovative project when you see it" has meant that the process of getting there is not well specified or explicitly taught. Connecting project elements against or across normative categorical barriers often happens later in the postfield analysis stage. But we believe that you can make these kinds of connections throughout all phases of the research process, including the very beginning phases. In this way, our handbook is both new and not new. It simply encourages what happens felicitously in outstanding ethnographic design, and it is dedicated to making that outcome accessible.

STUDENT COMMENTS

–   I just can't articulate here how helpful this course was to my project, and beyond that, my progress as a scholar. I came in with a swath of disorganized data and really no concept of how to work through it.

At the end of this course, I feel that I both have turned it into some-thing solid and have the skills to continue doing so over the course of my dissertation work and my career.

– This process takes what seems to be a mysterious, esoteric, or common-sense part of qualitative research—project conceptualization—and demystifies some of its most important components without over-simplifying the complexity of what goes into developing a research project. The assignments in this course have helped me completely rethink my thesis research and set an agenda for the next several years. I have gained skills in this course that I will use for the rest of my career.

Through this way of designing, we support the critical reworking of anthropology as an institutional discipline and social practice. To do so, we advocate for creatively open, intuitive, and collectively centered ethnographic design that allows for noninstitutionalized kinds of attending, knowing, and sensing.

We take inspiration here from other writings on "ways" of doing, fol-lowing the lead of other handbook writers who argue that transformative work requires deep attention to embodied preparation, rest, and collective spiritual attunement.[4] Julia Cameron's *The Artist's Way: The Spiritual Path to Higher Creativity* provided a strong guide.[5] Cameron's handbook provides twelve weeks of exercises to uncover and break through blocks to creative work of all kinds. It's inspiring because it insists that creativity and imagina-tion are inherent and that they require as much deliberate care and cultiva-tion as scientific knowledge production. We also are inspired by Felicia Rose Chavez's *The Anti-racist Writing Workshop: How to Decolonize the Creative Classroom*. In it, Chavez shows how to work against "traditions of dominance" in the classroom. In her workshops, she specifically acknowledges the affec-tive dimensions of those traditions, and she restructures learning and writing hierarchies to undermine normative and damaging patterns.

This handbook's multidimensional approach and emphasis on collec-tive process are also inspired by decades of changes in ethnographic practice. Anthropology's shifting terrains began when scholars in the late twentieth century insisted that fieldwork and writing processes were as political as theory and analysis. In particular, the 1980s writing culture debates were re-sponses to concerns about anthropology, representation, and imperialism.[6]

They surfaced the problem of ethnography's biased compositional and literary structure. Feminist, racial, and postcolonial analytical responses to these issues showed how the experience of ethnographic writing and reading is always situated in the gendered, racial, and class-based subjectivities of researchers and readers.[7] This prompted anthropologists to reflect on the production of research and theory responsive to emerging questions about the political nature of anthropology as a discipline and ethnography as a practice.[8] Furthermore, ethnographers began to draw attention to what "spaces" and "scales" meant in terms of examining the complexity of relational interconnection at large.[9] More recently, ethnographers concerned with methodology insist, in different and not always aligned ways, that fieldwork praxis and training must be attuned to emerging forms of research politics, ethics, philosophy, and creativity.[10]

The content of these debates and provocations is quite varied. Yet most (not all) assume one peculiar thing: that anthropological methods and theories are where disciplinary transformations begin. We amend this assumption. We assert that *research conceptualization and design* opens a foundationally powerful and vibrantly imaginative space for shifting disciplinary conceptualizations and practices. In fact, qualitative methods books have emphasized this point via calls for intersectional and anticolonial approaches to research design.[11]

When we teach and facilitate workshops on developing ethnographic research projects, we find that participants have relatively few difficulties with learning fieldwork methods, including addressing the ethics and politics of research interactions. Far more challenging for learners—and even for our experienced colleagues—is grappling with the slip-sliding problems of early-phase project planning within ever-changing research milieus. Problems that plague all of us include trying to engage newly urgent-seeming topics that are difficult to define, situating a project within overwhelming volumes of literature, managing fuzzy and untethered research concepts, ending up with elusive research questions that don't connect well to data-collection plans; and fielding uncertainties about engaging in relationships with people and places. Given these problems, it is difficult to stabilize a core cluster of researchable concepts and questions. As we read more or gather new data, clusters seem continually to disconnect or break down. Certainly, dealing with in-progress design changes and disintegration can be overwhelming; along the way we can feel anxiety about whether we are designing an intellectually significant and socially impactful project.

Therefore, this handbook consistently addresses the need for a project to be innovative and significant, as well as to be meaningfully connected to scholarly and broader communities. As such, careful attention to research design processes also has implications for research justice, in part by opening ways to come into community in the early imaginative and concept-work processes. While we provide only one among many possible ways to design ethnographic research in collective modes, we hope it might serve as a guide for cultivating and sharing other unique ways of making a project. In helping designers mindfully attend, with others, to the relational, political, and intentional dynamics of the project, the handbook can be useful in helping to establish ethical, reciprocal, and nonextractive commitments of their projects. However, we leave the specifics of collective work to the wisdom of designers, their advisers and colleagues, and the communities they are engaging as they deal with the many dimensions of fieldwork planning. Our process, therefore, offers ways to make visionary design decisions part of an ethically responsive and socially connected practice. This makes project development a mutually supportive rather than lone process, which helps produce projects that have conceptual and political integrity.

However, this goal is often difficult in the structural conditions of our working lives. Research design usually (but not always) takes place in academic environments that encourage conformity, competition, and individualism. In the United States, where we work and teach, these demands come from settler colonial institutionalizations that prescribe narrow and fixated approaches to research and what counts as "conceptual" or "empirical." As a result, there can be little room for openness, curiosity, play, speculation, flexibility, and thinking and relating otherwise. The *otherwise* in anthropological and other work is about being present to and materializing alternative social ways of interacting and being.[12] We hold that such endeavors are vital to becoming proficient in research design *as a peer-based collective craft*. We provide suggestions about how to work collectively with peers; we intend that these be modified to meet the researchers' own collective inquiry practices.

This book is for people at all levels of design experience: undergraduate students who are developing an ethnographic project for their methods courses or who are initiating an honors thesis or longer-term independent study; graduate students who are conceptualizing their projects; experienced researchers who want to refresh their approach to project construction; and those planning to teach research design.

While the handbook is designed for anthropologists conducting ethnographic work, it is also for those engaging in ethnographic and autoethnographic work across inter/disciplines such as sociology, geography, history, comparative literature, political science, creative writing, gender and sexuality studies, and ethnic studies, to name a few. For those conducting ethnography in ways other than it has been normatively defined, or even those who are *not* conducting ethnography, we invite you to hack our use of "ethnography" and substitute it with your inter/discipline's own key concepts and methods. Thus, the conceptualization and design process found in *The Ethnographer's Way* is applicable across multiple disciplines.

If you plan to teach this book or to use it to guide a group endeavor, we suggest doing all the modules yourself first. Not doing so would be like teaching a musical instrument without having learned to play it. That is, engaging this material requires an *embodied* sense of knowing the process. Even if you are a seasoned ethnographer, we think you will find, as others have, that this process will reorient and probably reenchant your project development experience. It tends to be a joy to teach because it empowers new and established researchers to unearth what is important to them while assembling a project with clearly evident intellectual and social significance. It can also help them get in touch with skills and knowing that they didn't realize they had, which can be healing for many who do this work.

As instructors, we understand that the multidimensional design process needs to unfold carefully or it could be overwhelming. A very common reaction to project overwhelm and unwieldiness is to cut elements and add new ones without having a way to attend to the whole project. In this handbook, we address how to make a project that can be altered as needed, in ways that maintain a flexibly integrated theoretical core. In addition, we offer ways to imaginatively expand and pragmatically contract the overall project scope—that is, its empirical and theoretical range—in effective ways that don't reproduce normative hierarchical ethnographic scopes like "local to global" or "home to nation." Ultimately, this kind of attentive work results in a conceptual assemblage that sustains congruence between the research objectives and data collection and is also exploratory and innovative. We get there by using techniques that engage minds, bodies, and collective energies.

This handbook can be used and modified (hacked) in or out of the classroom. As a standard institution-based teaching process, it can be used in a class that spans a quarter (ten weeks) or a semester (about fifteen weeks). In those teaching frameworks, it can be used to structure an entire course on design or used as part of a broader course on design and methods training. As a classroom guide, it is ideal for graduate-level training but can be adapted to give undergraduates experience in developing research topics, planning and executing research, and writing an ethnographic paper.

The handbook is devised to be used in workplace or workshop settings, where individuals or teams are preparing to conduct ethnographic research. This can include collaboration with interlocutors engaging in participatory and justice-based research design. We encourage those who are planning to use this handbook in the classroom or in a workgroup to consider modifications appropriate for their settings. For example, you might spend more time on modules or add exercises on actual methods to use and practice (for longer academic terms). Another option might be to break up the design work by allowing participants to field-test methods in tandem with the design exercises. Lastly, this handbook can also be used to support or clarify proposal designs and grant-writing processes.

After reading and planning for modifications, instructors and research collectives who plan to engage in group feedback should create an exercise submission plan for each module. That is, individual designers should complete all exercises in each module, but instructors or research collectives may want to specify which exercises they wish individuals to submit for coursework or group review.

There are two guiding process principles that we offer to both research designers and research design instructors: *valuing iteration* and *practicing in community*.

Valuing the Iterative Process over and above Instantly Materialized Results

In our teaching, we find that students attempt to power through and get things done quickly because that's what they're trained to do for all their classes. An enthusiastic attitude is a good disposition and can generate energetic activity. But yielding to anxiety-based pressures to get things done and dusted can be problematic when performativity and competition are the driving cultural forces of the work. This can deaden curiosity and the capacity to carefully reorganize one's work in innovative and liberatory ways.

Our efforts to pace this handbook to stay process-oriented and provide breathing room, even within the structure of academic terms, are centered on student-researcher well-being. But we also want to provide a rewarding and illuminating experience for researchers and instructors as they open new possibilities for research design interactions. We want both students and instructors to recognize that an unrelenting "powering through" and "finish-it-up" attitude can be harmful to experienced and novice researchers alike. We've seen, and personally felt, such embodiments lead to self-doubt, chronic avoidance, and other threats to well-being.

We also recognize, however, that efforts to shift out of "results now!" paradigms can be rattling for learners, instructors, and others, so we address this problem in our modules. It's difficult to trust a process whose outcomes will materialize in good time and often at different paces. We think this needs to be openly managed in the classroom and workplace. Being clear about the emotional highs and lows that accompany this work helps to regulate emotional reactivity (in order to remove obstacles to intellectual insights), encourages self-trust (which diffuses self-doubt), and normalizes a cooperative sympathetic joy for all intellectual pursuits (which can diminish the competition and aggression that we are all socialized into from a young age).

To counter worries about the lack of instant certainty, we remind novice researchers in our courses that they may not get to their final multidimensional design in ten weeks. Instead, we emphasize that the focus and aim of the course are on listening to the project and gaining conceptual project design skills. These skills are designed to become intuitive, which is the aim of any craft. And so when students finish our course, they will have learned a lot about their projects but not everything. They leave knowing that they have the know-how to help the rest of the project conceptually unfold long into the future.

Lastly, we encourage instructors and mentors to be mindful of their own professional urgencies and expectations. Without meaning to, instructors and mentors often want to help learners toward finality in ways that don't always cultivate enough space for intuition, listening, staying open, and remaining curious. We advise that breakdown and slowdown in the design process not be pathologized by instructors.[13] Research designers need to be told this and guided through conceptual challenges—challenges that are usually just typical encounters with designing a long-term research project. Such moments are often encouraging signs that the project is in process, and it may need a focused revision or more radical forms of letting go. That is, breakdowns and slowdowns can signal attachments that prevent one from going back to earlier steps or from experimenting with alternative design elements.

Instructors can adapt our exercises to provide wise guidance as learners move through design ups and downs. They can, for example, allow learners to move through their uniquely powerful responses to the uncertainties of research design, including making space for expressions of self-conscious hesitancy and personal or communal trauma. In practical terms, instructors and mentors might suggest that learners take a process break, redo earlier exercises, or do some free-writing exercises to get unstuck and centered again. When instructors prioritize processes of allowing and guiding together, we find that learners arrive at a project that "feels right" and that is also intellectually and socially significant.

Practicing Community Necessity

We feel it's important for learners to be effectively supported by working within a community of peers every step of the way. Community isn't an option: it is a necessity. For this process, we address how to establish and practice community necessity in more depth in Interlude I. At the end of each module—in the "Collective Concept Workspace" sections—we provide prompts that (1) emphasize the iterative aspect of design work, for the purpose of (2) cultivating a strong intuition for organizing multidimensional projects that (3) slows down the process so that openings to project insights can be possible.

Such insights flourish in a consistent, supportive environment provided by the communities you create. We recognize that finding community can be easy for some and difficult for others. In academic settings, students can feel marginalized in their programs. But in the same way that we may choose our "families" and friends, community here means finding ways to choose our allies and in/formal mentors.[14]

The necessity of intentional collective design work leads to an enacted politics of support that gets us out of the kind of individualism that we are enticed by and awarded for in institutionalized professions. We advocate breaking the interior and exterior bonds of that institutionalization. We all need different places and paths for creating knowledge. The scope of contemporary political problems calls for a radical rethinking of our professional patterns. Working within our own community-created processes provides one avenue for doing research otherwise. If there is momentum for the kind of practice we offer, then we hope that structures that foster exclusion and competition will give way to more liberating ways of being in our research and academic worlds together.

ACKNOWLEDGMENTS

When we first started working in the Anthropology Department at University of California Irvine (UCI) in the late 2000s, the graduate training program's second-year required methods sequence consisted of two quantitatively focused, ten-week-long methods courses. These courses were developed by the department's founding anthropologists to advance mathematical models of social behavior, measurements of cultural knowledge, and quantitative multidimensional scaling. (The latter was developed by A. Kimball Romney and not to be conflated with our multidimensioning approaches.) We are grateful to Michael Burton and Bill Maurer, who dedicated themselves to transforming these courses into fascinating and impactful science and technology studies approaches to anthropological methods.

By 2013, Mike retired, Bill became dean of the School of Social Sciences, and George Marcus became chair of the department. George approached us and asked if we wanted to "play around" with these courses. We enthusiastically accepted and, along with other colleagues, embarked on transforming the second-year sequence from two quantitative methods courses to three qualitative project training classes: methods, research design, and grant writing. At first we spread our conceptual approach across the anthropological methods and research design courses. Now we teach the entirety of this handbook in the ten-week Research Design course.

It took many years and lots of experimentation before we produced *The Ethnographer's Way*. We would very much like to thank George Marcus for trusting us with core graduate training. He continues to be a generous colleague and friend, as well as a generative interlocutor for our research design endeavors. As graduate students at Rice University, we worked closely with George as well as Jim Faubion, Nia Georges, Chris Kelty, Hannah Landecker, Ben Lee, Amy Ninetto, and Julie Taylor. We deeply thank them for their support in a creative environment that shaped our conceptual thinking as anthropologists, research designers, and teachers. After we finished at Rice, Kim

Fortun, also a Rice alum, generously shared her graduate ethnography syllabus with Kris in 2006, which, importantly, showcased Kim's innovative practices in the classroom that have evolved into her own conceptual training.

We are deeply grateful to the UCI Anthropology faculty who may have often wondered what we were up to—especially as we came up with strange new project design terms to use with students. But they nevertheless had faith and encouraged our experimental endeavors. We especially acknowledge our colleagues who are also instructors of the second-year sequence who, over the years, have supported our and each other's teaching endeavors. Much gratitude to Victoria Bernal, Tom Boellstorff, Kim Fortun, Sherine Hamdy, and Eleana Kim. We thank all of our colleagues for their kindness, support, and inspiring teaching and scholarship: Samar Al-Bulushi, Mike Burton, Leo Chavez, Ricky Crano, Tom Douglas, Chris Drover, Jim Egan, Julia Elyachar, Mike Fortun, Angela Garcia, David Theo Goldberg, Anneeth Kaur Hundle, Angela Jenks, Karen Leonard, Chris Lowman, Lilith Mahmud, Bill Maurer, Keith Murphy, Sylvia Nam, Sheila O'Rourke, Justin Richland, Taylor Riley, Damien Sojoyner, Ian Straughn, António Tomás, Roxanne Varzi, Sal Zárate, and Mei Zhan.

Other colleagues have read, taught, or otherwise engaged this work and provided invaluable feedback, great dialogue, and much kindness over the years. Many thanks to Samar al-Bulushi, Kamari Clarke, Jatin Dua, Joe Dumit, Sahana Ghosh, Anneeth Kuar Hundle, Kaushik Sunder Rajan, Negar Razavi, Emilia Sanabria, Gabrielle Schwab, Damien Sojoyner, and Madiha Tahir. We are grateful to Ellen Kladky for expert copyediting!

We also thank Teresa Montoya, Natacha Nsabimana, and the Anthropology Department at the University of Chicago, who hosted Valerie for a talk in 2021. Students especially asked questions that helped transform the names of our core concepts, which, at the time, represented more hierarchical and linear thinking—what we were trying to get away from. Tim Choy at UC Davis hosted Kris (along with Kaushik Sunder Rajan) for a talk as well as extensive day-long conversations with Marisol de la Cadena, Joe Dumit, Alan Klima, Fatimah Mojadeddi, James Smith, Smriti Srinivas, as well as many UC Davis graduate students. These dialogues eventually led to new experimental class exercises that now appear in the book.

Some of the deepest thanks goes to Emilia Sanabria who organized a 2.5-day multidimensional design workshop at a retreat center outside of Paris. She did so without knowing the full extent of what we were doing. We deeply thank the participants at that workshop: Claire Beaudevin, Charlotte

Brives, Fanny Chabol, Denis Chartier, Joe Dumit, Chris Elcock, Sophie Houdart, Sylvia Mesturini, Mariana Rios, Emilia Sanabria, and Piera Talin. They helped us with profound conceptual insights regarding the corporeal and somatic aspects of research design. It was a high-energy, ground-shifting time for all of us. To this day, we are truly grateful for such a rare experience that most of us never had at an academic event. After that workshop, we've had the privilege of ongoing, insightful, and brilliant engagements and observations with Joe Dumit and Emilia Sanabria. Valerie also thanks her colleagues Daniel Knight, Valentina Marcheselli, Perig Pitrou, and Istvan Praet who allowed her to integrate some of this material into invited talks and conversations in Europe and the United Kingdom. Kris additionally thanks Hannah Appel, Sean Brotherton, Larisa Castillo, Mǫ́rẹ́nikẹ́ Foláyan, Angela Garcia, Avery Gordon, Laura Johnson, Julie Livingston, Metsi Makhetha, Sabrina McCorkmick, HLT Quan, Nonie Reyes, Elizabeth Robinson, and Judy Talaugon for their friendship and intellectual kindness. We have so much gratitude for our colleagues's friendship and support.

Most of all, we thank all the students who took, and will take, our courses on design and analysis and to whom we dedicate the development of this process. They are always open to experimental classroom practices as well as giving us room to name the moments that we didn't know what we were doing. They provide a great deal of patience and enthusiasm. Every week we learn from their research and, more importantly, learn how to better conceptualize and teach this work.

This book would not be possible without these students. Sincere and enthusiastic thanks goes to Gray Abarca, Ahmed Adam, James Adams, Tawfiq Alhamedi, Chima Anyadike-Danes, Kyrstin Mallon Andrews, Lucy Carrillo Arciniega, Monique Azzara, Nandita Badami, Shannon Bae, Nicola Bagic, Nataly Bautista, Melissa Begey, Michael Briante, Emily Brooks, Colin Cahill, Alice Chen, Alexandra Chmieleswki, Joshua Clark, Nathan Coben, Evan Conaway, Benjamin Cox, Katie Cox, Charlie Curtis, Marc Da Costa, Liz DeLuca, Upuli DeSilva, Cheryl Deutsch, Nan Ding, Nathan Dobson, Garrison Doreck, Emily Earl, Raymond Fang, Nasim Fekrat, Margaux Fisher, Akil Fletcher, Colin Ford, Diana Gamez, Melissa Gang, Lucy Garbett, Vida Garcia, Anissa Gastalum, Oviya Govindan, Padma Govindan, Courtney Graves, Mariel Gruszko, Kelly Hacker, Gina Hakim, Georgia Hartman, Tim Hartshorn, Forest Haven, Muneira Hoballah, Tannya Islas, Emily Jorgenson, Scott Jung, Anna Kamanzi, Neil Kaplan Kelly, Robert Kett, Christine Kim, Hae Sue Kim, Inah Kim, Mette Kim-Larsen, Ellen Kladky, Alex Knoepfelmacher, Greg Kohler, Yimin Lai, Matthew

Lane, Orlando Lara, Juwon Lee, Janelle Levy, Neak Loucks, Guilberly Louissaint, S. Zaynab Mahmood, Sean Mallin, Emily Matteson, Andrew McGrath, Kimberley D. McKinson, Colin McLaughlin-Alcock, Chandra Middleton, Tarek Mohammed, Alberto Morales, Rojelio Muñoz, Megan Neal, Angela Okune, Adan Martinez Ordaz, Jason Palmer, Justin Perez, René Perez, Simone Popperl, Matthew Porter, Farah Qureshi, Kaitlyn Rabach, Tariq Rahman, Lili Ramirez, Muhammad Raqib, Beth Reddy, Rebecca Richart, Darwin Rodriguez, Liz Rubio, Katherine Sacco, Camille Samuels, Daina Sanchez, Leah Sanchez, Linda Sanchez, Rosie Sanchez, Tim Schuetz, Nick Seaver, Taylor Silverman, Jessica Slattery, Isabelle Soifer, Prerna Srigyan, Danielle Tassara, Mindy Tauberg, Michael Tecson, Heather Thomas, Natali Valdez, Yvette Vasquez, Tenzing Wangdak, Sandy Wenger, Josef Wieland, Annie Wilkinson, Alex Wolff, Melissa Wrapp, Fei Yuan, Nima Yolmo, Leah Zani, and Jennifer Zelnick.

With much gratitude, we thank students who cotaught our undergraduate version of this course. These experiences helped refine what we were teaching in the graduate seminar. With each iteration, we all agreed that teaching project design and guiding undergraduates to analyze data, and produce papers and films out of the process, were some of the best mentoring experiences of our lives. Deepest thanks to Colin McLaughlin-Alcock, Mariel Gruzsko, Tim Hartshorn, and S. Zaynab Mahmood. Valerie also thanks her postdocs Ségolène Guinard and Michael Vine.

We have so much gratitude for former students (most now working in various professions) who were in our courses and who agreed to let us use their in-progress work as exercise examples: Nandita Badami, Forest Haven, Ellen Kladky, Sean Mallin, Kimberley D. McKinson, Colin McLaughlin-Alcock, Jason Palmer, Tariq Rahman, and Annie Wilkinson. We are very inspired by their compelling research and savvy multidimensional thinking and teaching.

We are so happy that Ken Wissoker at Duke University Press was supportive and enthusiastic about this project. Deepest appreciation for you, Ken! We also thank all the people and teams at Duke who made this handbook come into being, including A. Mattson Gallagher, Kate Mullen, David Rainey, Christopher Robinson, Laura Sell, and most especially Bird Williams, our project editor.

We are both joyfully indebted to our families and life mentors who supported our work on this handbook through the pandemic and our illnesses. Valerie wishes to thank her partner Guy; her daughter Morrigan; her mother, Gail; her sister Hillary, her brother Kurt, and her godmother, Phyllis. Her friend and colleague Roberta Raffaeta provided steadfast companionship for thinking across dimensions over five years. Kris thanks her mother, Suzie; and siblings Susie, Todd, John, and Drew, and their partners and kids; she writes in memory of Robert Simon. We are indebted to our teachers and guides: Kris Abrams, Shannon Dailey, Marianne D'amore, Clara Favale, Nancy Guiliani, Bruce Johnston, Melissa LaFlamme, Jessica Mathon, Dave Smith. Additionally, we thank our oncologist, Robert Bristow, and the health teams at UC Irvine Medical Center who provided care and support while writing this book. We hold many people in our hearts who inspire us. We are grateful that they have shown us that practicing the spirit of generosity, nourishing kindness, and collective liberation is meant for all who wish to live and love deeply in our worlds.

Multidimensional Concept Work

In this handbook, we show you how to design an innovative and socially responsive research project. Throughout the process, we invite you to keep a beginner's mind. This asks you to stay open to states of *not*-knowing and to welcome the process of learning with others. This invitation makes the handbook more than a toolkit. It can become a way to integrate research's personal and collective possibilities.

We have been helping people experience the joys and challenges of research design for years, cultivating activities that result in well-integrated and successful projects. We take you through these activities step-by-step. After you complete the handbook, you will have all the elements of a coherently assembled project: a sound integration of relevant literatures; a compelling description of your topic and aims; a coherent theoretical and empirical object of study; and elegant research questions that shape the project's scope, data sets, and field interactions. You will need to be familiar with ethnographic methods and ethics in order to complete the module on field interactions, but this handbook will help you produce all the other elements. Along the way you will put these elements into a cogent project grid, which provides the basis, later, for writing a strongly congruent research proposal.[1] All these elements are standard for a successful research design process. But our process is also *non*standard in that it also includes ways to help you attune these activities with your existing experiences, intuitive intentions, creative activities, and collectives.

Our aim, overall, is to make research design a transformative process—meaning one that does more than produce a proposal or thesis. We hold that ethnographic research design, with its processes of imagining and engaging, offers opportunities to connect knowledge production processes with transformative practices in intellectual work and perhaps even social change. In this way, we advocate pushing the institutional boundaries of anthropological inquiry so that researchers can better align academic projects with broader values and goals. In short, we hope to make research design a process with scholarly as well as visionary dimensions.

This integrative approach treats projects as intellectually and personally multidimensional. Let's walk through a definition of our design approach, and then we'll detail what makes a research project *multidimensional* and how *multidimensioning* works as a design technique.

Multidimensional research design is an iterative approach to assembling diverse research concepts and intentions within a congruent framework of inquiry.

We use the word *concept* for all the terms that constitute your project. The concepts you will assemble in this handbook are diverse: empirically specific as well as general and theoretical. They may be rendered in the language you are using to develop the project but also could include terms relevant to those with whom you'll be working. Concepts include varieties of beings, objects, places, processes, and contexts you will directly encounter as well as theoretical ideas and other creative elements you want to bring into your design. In other words, concepts specific to a project's particular beings, things, activities, and places like "land," *Haa Atxaayí Haa Kusteeyíx Sitee,* "digital deeds," "food," *mohallas,* "saving money," "fake drugs," "Peru," and "Alaska," as well as broader contextual or processural experiential concepts like "love," "racial enclosure," "neoliberalism," "financialization," "security systems," *rizq,* "liberation," and "democracy." You will find these concepts in this handbook's examples. You will derive your project's key concepts from mapping out your own knowledge and intuitions and from doing literature reviews (in Modules 1 and 2).

We emphasize that the key concepts you decide to engage in your project are interdependent and contingent. Each concept must be understood to be in relation to the others. Understanding concepts as modular terms in this way helps you change or replace them if needed. Keeping modular helps you maintain curiosity and respect for what emerges when you assemble different concepts in different ways—what we will refer to as *concept combos* (Module 4). For example, putting "farmers" and "Argentina" and "soy" together in a project dislodges a set meaning for any of these stand-alone terms. To farm soy, rather than something else, in Rosario, Argentina, is a unique process, and it is not the same as farming soy in Maharashtra, India. When you regard terms as flexible concepts that change in relation, you can keep their contingent concreteness and abstractness in view. This helps you decide how to select and relate concepts in intuitive and newly compelling ways.

The *iterative* aspect of the multidimensional design occurs as you work forward and backward at the same time via practices of finding, assembling, reflecting, revising, and iterating concepts. This helps you create conceptual connections that are meaningful but not formulaic or hierarchical. We help you to cohere these conceptual connections to create a *congruent* ethnographic project that also integrates other concerns you have—social and political—that extend beyond the practice of ethnography. In the final phase of the handbook's process, you will produce a congruent *framework of inquiry* with broad theoretical and specific data-collection questions that are inspiring and answerable. Given that your project design clearly demonstrates a relationship between the gaps you found in the existing literatures and your data-collection plan, your answers to those research questions promise to be intellectually and socially significant.

We provided one example of a multidimensional project with these features in the prelude; here is another example we expand on in this handbook. Forest Haven, a researcher we worked with, designed a project that created a new conceptual framework for examining the politics of Indigenous food sovereignty in Alaska. She could have simply focused on Native food as an ingestible *material dimension* of life. But based on her personal and field experiences, she intuited that she could also examine how different groups of people experience food as a nonmetabolic and *sensorially embodied and technical dimension* found within unique (Alaskan) Natively and colonially defined spaces.

The result of putting those seemingly disparate dimensions of food together is a *multidimensional* project that dynamically combines two conceptual dimensions: the environmental and the sensorial aspects of Alaskan food sovereignty politics. The project does so within a broadly spatial and temporal colonial context. Haven's literature review of this conceptual combination, a vital part of the multidimensioning process, confirmed that this was a productive and underexamined ethnographic concept combination throughout literatures on food politics in general.

Let's now take this moment to discern the difference between *multidimensionality* and *unidimensionality*. Haven might have designed a conventionally legible single-dimension project by focusing just on "Alaskan Indigenous food consumption and procurement in a changing environment" rather than focusing equally on the "sensing" dimension. Such a project would have examined what is locationally and culturally unique about Alaskan food: an edible and gatherable substance in the context of contemporary environmental degradation and food insecurity. However, at the conceptual level this creates only one-way dimensioning, meaning that people (Alaskan Indigenous people) and food-as-matter-in-environment are only indexing each other in the design framework. This could limit the researcher to designing a standard "food case study" form of ethnographic theorizing. Such a project wouldn't have reason to reach further—to plan for fieldwork on how food impacts the social and political dimensions of life as something in relation to spatial belonging and perceiving in other ways than eating and gathering. Nor would it be able to account for, or theorize, the practices and effects of Alaskan settler colonialism as a way to control environmental embodiment. Such a case study project certainly might contribute to a geographically situated "food and peoples" gap in the literature. But it wouldn't "pull" the project "outward" into the deeper conceptual combinatory multidimensionality that emerges by combining eating and gathering with Indigenous and state processes of sensing.

In contrast, Haven's multidimensional approach reimagines her project's theoretical and political possibilities—and what food is as an aspect of lived, spatialized experience. Based on her conviction that food sovereignty is intimately related to processes that make food uniquely sensed in Alaska, by both Native bodies and colonial surveillance technologies, she intuitively juxtaposed concepts of seemingly different kinds to open up and/or reframe her project questions. In particular, she was able to interrelate a chronically underexamined dimension of "Alaskan Indigenous food"—the broader embodied and technical dimensions of "sensing" with the broader contextual dimensions of "colonialism" and "sovereignty." In one of her project narratives, she explains, "I am juxtaposing sensing and sovereignty in order to understand the meaning of food governance and dispossession on Native terms; I am studying how sensory politics are not 'local' or 'personal' bodily politics, but relate to larger spatial schemas of collective embodied sovereignty and conflict."

When the ethnographic and theoretical dimensions of food, sensing, and sovereignty are put together, new theory-making and social activist possibilities arise; these can include new or alternative ways to understand how modes of colonial governance, dispossession, and sovereignty emerge through very different processes of sensing: that of Native communities and that of the state.

Multidimensionality, therefore, isn't about random conceptual aspects jammed together; rather, it's a small, well-curated set of conceptual combinations that pulls the research into new possibilities. What makes such a project innovative is that it has a unique conceptual shape, helping us engage social processes differently.

Here is a definition of the design technique of multidimensioning. Come back to it when you need a design inspiration refresher.

Multidimensioning is the process of defining a project's conceptual combinations and using them to create congruently integrated project elements, from research topic to research questions.

Multidimensional research design addresses the need for ambitious research in urgent and complex political times. The entangled dynamics of racism, climate change, pandemics, indefinite war, precarious life, and volatile economies, to name a few, coexist with rising forms of resistance and otherwise ways to be and inhabit.[2] Otherwise ways of conceptualizing and doing creative work can get us out of modernity's linear, hierarchical, and racial prescriptions. Such ways out are not individualistic; they require mutuality. As Emilia Sanabria writes, "The taking care, the dwelling in, and staying with the trouble throughout is what makes the otherwise possible; it includes a simultaneous dynamic of struggle and collective imagination that is not just *against* but powerfully oriented *towards* something otherwise."[3] Today's ethnography, then, has the potential to engage the multiple dimensions of distinctive and interconnected lifeworlds.

In what follows, we cover the "connective tissue" components of our particular approach: multidimensional concept work, imagining, project listening, and working in community. We encourage you to continue to adapt it for yourself and with others.

Designing with Connectivity, Intuition, Curiosity, and Congruence

To embark on this way, you will engage with others in flexible but disciplined *concept work*. In anthropology and other vocations, concept work generally refers to connecting ideas and things in order to form useful working constructs that lead to generative questions and fieldwork. In research design, concept-work "think tools" help you construct a project.

We understand concept work to be a fully embodied activity, and for this reason we address the importance of care, pausing, and reflection on a personal and collective level. This handbook's concept work aims to nurture your intuition and imagination, generate and test research feasibility, stay true to your curiosity and intentions, and manifest project congruence and potential.

Multidimensional concept work encourages an intuitively attuned approach to connecting project elements in ways that feel, in your body and in your social experiences, true to life. By intuition, we mean those inspired flashes of whispery inklings that don't necessarily arrive from normative reasoning processes. Listening to these intuitive inklings enhances your capacity to reflect on what you think you know, what you don't know, and what might be productively counter to standard notions of how things relate (for example, relating the process of sensing food with eating food, or of having a soul with seeking citizenship). We give you techniques to explore such experience-based intuitions that "this might connect to that," whether you are new to research design or not. Intuition is essential for honing any craft such as art, architecture, carpentry, cooking, making music, and so on. It helps a creative worker to dream something new into being and to explore and revise this living creation until it manifests their intentions in the best possible way.

We cultivate your intuition, with its inchoate states of interconnected feeling-knowing, while recognizing that in Western institutions intuition is only provisionally associated with intellectual work yet is entirely essential to it. We also know that the politics of intuition are deeply fraught on many levels. Conventional academic infrastructures can divert us from playing with our open-ended intuitions, especially when they reinforce hegemonic forms of institutional and imperialist performativity of "reason." Intuitive thought is also constrained by "grind culture" that perpetuates capitalism's breakneck speed and chronic neglect of bodies and minds, denying us, among other vital needs, the care, stillness, and rest that nurture visionary work. Yet intuition is an embodied mode of open-minded perceiving that's always present, always dynamic, and can be carefully developed. We believe that cultivating processes of slowing down, listening, and intuiting can increase the potential of anthropological projects to usher us lightly into other possibilities of living and relating.

Ethnographers often put intuitive creativity on hold until after data collection. But we don't want you to wait that long! We find that such intuitive leaps can be productive at the earliest phases of research design, especially when you think them through with others. So, we developed exercises and feedback processes that incorporate both intuitive and classic empirical modes.

In addition to supporting intuition, we also want to counter disjunctive institutionalized processes that kill *curiosity*. The vitality of an ethnographic project anchors in its accounts of what research already shows, but it also faces what is truly unknown. And, even more importantly, it faces what the researcher does not know. Unfortunately, the neoliberal academy is moving further from open-ended curious inquiry and deeper into instrumentalized research that can be "applied" in order to refine, reform, or rebrand existing structures. We have noticed that, in this milieu, students often anxiously construct projects that safely reproduce established work. They feel disempowered to explore what they, and their mentors, don't know about social worlds. They feel instead that they must fit their project into inquiry slots that are already deeply grooved, so they can demonstrate how their project aligns with given conclusions about everything from social cohesion and belonging to racism and capitalism. They hesitate to break out and do something different. For example, they might intuit that there are ways to ask new, perhaps unusual, questions—such as to put questions of economy together with cosmological or spiritual questions, or to ask about the colonial state in relation to questions about how people smell and taste and sense things—but they don't feel permitted to do so. The coming modules show how successfully intuitive and curiosity-based projects came into being.

In cultivating your intuition and curiosity, we aim to reempower your right to not-know and to join with people taking risks to imagine something new and open to difference. With all of this in view, we offer practices to cultivate the researcher's curious and intuitive body, heart, and mind in order to align not-knowing with imaginative possibility. Curious, intuitive, and imaginative practices can connect the project with liberative currents. And beyond producing projects, those practices can also support nurturing kindness toward our own and others' research processes. This leads the way to deep insights and care for our intellectual communities.

In sum, effective concept work enriched by imagination and curiosity results in a project that stays flexibly congruent at every stage of the research process. Although its concepts are diverse and connected in ways that may not seem coherent in a normative Western sense, the project hangs together as a whole from initial design to final writeup. This requires aligning the disparate aspects of research design—people, sites, things, theories, literatures, methods, processes—within a flexible but harmonious framework.

To maintain this congruence, researchers need to clarify and re-clarify project elements and their connections throughout project design, research, and writing. That is, a project requires *iterative* redesign across its life course. This handbook has ten modules of proven concept-work exercises that address these and other design challenges; in Module 10, we explain how to adapt these tools to produce analyses and other final products. The sequence unfolds as it does to save you lots of trial and error, but also allows you to double back and redo exercises when necessary. The first process that we teach you will help you address how to connect obvious but also intuitive inklings about what belongs in the study.

Beginning with the Research Imaginary

We usually begin a research project with the intent to ask and answer questions in ways that generate powerful social critiques. Critique is, of course, a necessary political and revelatory act. It is at the heart of any social science training. Effective, engaged critical analysis is refined over intellectual lifetimes. But critiquing is often seen as more rigorous or impactful than imagining. In contrast, in this handbook, *imagining* is a foundational design practice meant for empowering curiosity and engaging otherwise visions. In *The Ethnographer's Way* we enact our conviction that creating a critically attuned, socially responsive framework of inquiry requires imaginative stretching and experimentation. Critical thought does not have to be separated from imagining new ways to conceptualize projects and to be in worlds together. Understanding why this is important in the research design process requires a deeper look at the historical dissociation of thinking and imagining.

The history of social science is rife with convictions that critical reasoning is a higher perceptual form than intuition or imagining. This conventional institutional attitude (per Aristotle's *De Anima*) holds that imagining is a sensation-based process of thinking in nonlogos forms, like images, that must be disciplined through linear processes like narration and formalized logic. The implication is that imagining simply indulges in messing around with ideas about existing social forms in undisciplined and frivolous ways, as opposed to reasoning about them in a categorically driven, linear, and logically orderly fashion. According to this perspective, what someone imagines, therefore, is perceived but in an ultimately disorderly and irreal way. Imagination, in this perspective, can be used to stir new ideas, but only if those ideas can be symbolically ordered.[4] These perspectives bifurcate conceptualization into two streams: those that are orderly and rational, and others that are free-form or nonlinear. That is, imagination is deemed a place of fantasy, rather than an avenue of insight or perception. In contrast, we invite you to see how imagination can recognize the *real* nonlinear and unordered connectivities of living experience that critical reasoning, with its preordained categories, hierarchies, and scales, can overlook.

When we use *imaginary* as a noun to refer to making a written account of what you imagine—as we do in this handbook from Module 1 onward—we are deliberately breaking down irreal boundaries. We want to free researchers to constellate what they imagine to be probable and possible in ways that aren't linear, in service of narration, or in service to standard forms of generalizing or realizing.

By taking the imaginary as a fully-fledged form to design with, we follow theorists such as Sylvia Wynter and Báyò Akómoláfé, among others. These theorists hold that given Western conceits about the need to segregate sensing, feeling, intuiting, and imagining from disciplined or scientific thinking reproduce body/mind splits within individuals and across race and reinforce liberal mythologies about the ascendance of conventionally rational Man.[5] This mandate to segregate and order perception can extend into anthropology's research design practices and epistemological frameworks, which operate almost by default, even as anthropologists consciously try to refute them.[6] As Édouard Glissant shows, Western modes of splitting and reductionism force our very humanity into foreclosed modes of being and knowing, such that we do not even possess the everyday language to inquire and imagine what life could be like outside of these reductions.[7]

Therefore, by encouraging your capacity to intuit and imagine, we hope to walk with you out of reductionist liberal and positivist frameworks and into new epistemological and practical possibilities for your projects and what you can do with them in worlds.

Fortunately, anthropology has traditions and emerging practices that honor curiosity, intuition, and imagination. *Imagining, in such practices, is an active process that requires cultivation and faith in otherwise creativity and thought.* Our design method is, for example, inspired by Zora Neale Hurston's insistence on pairing imaginative storytelling with ethnographic reportage, which connects imagining and analyzing.[8] It is thus in alignment with scholars who explicitly link imagination with social analytic processes. Some of our exercises reflect C. Wright Mills's admonishment to turn comfortably obvious research ideas and questions on their heads and upside down in order to open up the unknown.[9] Other exercises reflect our solidarity with colleagues working to remake research methods and foster collective methods development.[10] Scholars pursuing such exploratory collective processes in academic and para-academic settings—such as Keith Murphy and George Marcus's ethnocharette, Andrea Ballestero and Brit Ross Winthereik's studio process, the Center for Imaginative Ethnography's play outside of the usual ethnographic boundaries, and the multimodal teamwork of Coleman Nye and Sherine Hamdy—advocate for collectively engaged iterations of reimagining, redesigning, and reanalyzing.[11] Many of these new processes are informed by the political conviction that otherwise ways of knowing and being stand outside and can act to dismantle liberal suppositions about what counts as an acceptable form of scientific or analytic creativity. Imagining, therefore, is not something that only proceeds research. It can also guide and shape it, resulting in tangible written and graphical works that are imaginaries in their own right, not simply the products of earlier imagination.

The imaginative concept work you will do in this handbook starts with Module 1, "Imagine the Research." From this module forward, we enact our support of other scholars who aim to conjoin research with counterinstitutional liberatory practices. We are especially inspired by studies found in the Black radical tradition pertaining to marronage and fugitivity, feminist approaches to experientially integrated and embodied modes of anthropological research development, studies of Indigenous political cultures as models for social relations as well as political organization, and multispecies work that engages post- and nonhuman subjectivities and more-than-terran worlds.[12] Following the leads that arise in these bodies of work we ask: *What if the purpose of curious and intuitive anthropological imagining is to let go of liberal segregations and reductions and create spaciousness for otherwise ways of inquiry and being to present themselves*?

Module 1 allows you to play freely with intuitions and ideas, helping you begin outside of patterned conceptual categories and topics so that your project doesn't come into view eclipsed by academic shadows of the past. We'll get to literature reviewing in Module 2, but first we want you to *reclaim the imaginary* as a space for experimenting with research possibilities and dimensions. In it, we provide exercises to let go of default prescriptions for "defining the field" so that you can engage in expansive but mindfully stepwise explorations of what is possible for your project. This first step is also where you'll start to practice the iteration and reflection you'll do from then on. Often, speculating and trying things out means that one's first leanings and intuitions are opaque and mysterious. Or they feel meaningful but it's not quite clear how they connect.

Imagining and experimenting with the elements of a project-in-process takes time. But the results can be groundbreaking—an outcome that is often best achieved in a collective space. You'll have lots of chances to imagine and reimagine, and in the end, you'll have imagined a project into view.

The imaginative and iteratively productive process of multidimensional design allows you to invest fully in what is unfolding and not just in end products.[13] Leaning into process with curiosity helps to keep Courtney Morris's question to anthropologists at the forefront: "What if we did the work as though our lives and the lives of others depended on it?"[14] This is not just a call to imagine outside of the box, but to leap out of it as Frantz Fanon invokes when he writes, "I should constantly remind myself that the real leap consists in introducing invention into existence."[15] Thinking processually with Fanon may include "reconfiguring the conditions of (our) own inner lives, reconfiguring the conditions of (our) immediate surroundings, and ultimately, reconfiguring the conditions of how we live our societies on Earth."[16]

adrienne marie brown and Walidah Imarisha underscore the far-reaching revolutionary value of imagining when they write that "the decolonization of the imagination is the most dangerous and subversive form there is: for it is where all other forms of decolonization are born. Once the imagination is unshackled, liberation is limitless."[17] And so it is in the spirit of this deeper vision and call to manifest what is possible that we approach multidimensional design as a mode, based on nonlinear, nonhierarchical interrelational practice.

Getting out of Vertical Scaling and into Multidimensional Space

In this handbook, we work in a multidimensional space; we view this as an alternative to staying within what we call the "vertical scaling" of normative research design. The Western "scale" concept is loaded with normative histories and connotations. Standard positivist "scaling" takes relative dimensions of things, such as color or size or distance, and grades them into hierarchically and sequentially ordered ranges, such as "low to high" or "near and far." These graded ranges can transmit dominant logics of *inherent* relative value, such as when "low" is subjugated to "high" or "near" is designated as more real than "far." The term can also reify hegemonic assumptions about *intrinsic* scalar difference, such as when the category of time is scaled into past-to-future and the category of space is scaled into near-to-far, and the two categories are kept separate. The precept that scales are inherently and intrinsically ordered and separate invokes images of ranked transcendence, such as that of an arithmetic progression, racial description, ladder, or tower. As a result, standard scaling can reify low-high hegemonic notions of difference, development, and value. Here we discuss how vertical scaling works and what multidimensioning offers otherwise.

In research design, researchers who approach design in terms of vertical scaling can end up framing sites and things within linear scalar orders with binarized poles. To name just a few of these orderings: standardized spatial schemas like local/global or home/nation; hierarchical forms of divided ontologies, such as that of body/society; progressive orderings of conceptualization that separate materiality from cognition, such as fieldwork/analysis or people/ideas; and the authoritative separation of theory-making from experiential learning, which actually depend on each other for knowledge production.[18]

We describe these standard linear scalar schemas as "vertical" because they generally reflect Western models of lowness-to-highness. Scholars and critics have used other terms to describe how this form of ordering is culturally and politically situated. Social scientists point to the historical construction of Western scalarity and its incommensurability with diverse nonlinear and nonmonocentric processes of relating and living. Feminist and feminist-Indigenous anthropological work, in particular, opens perspectives on the shape and extent of vertical scaling. Marilyn Strathern refers to the "holographic" nature of things-in-relation that do not map onto the West's discrete linear scales. Anna Tsing describes how the amorphous meshworked fields of organisms like mushrooms-in-relation-with-other-species cannot fit into the precisely hierarchized scales of Western biology and capital. Attending to the unbounded spatialities of more-than-human and more-than-terrestrial spaces, Zoe Todd writes about how Indigenous understandings of fish lives are based on perceiving animals as multifaceted "active sites of engagement" within multitudes of relations rather than as separate individuals existing along linear geospatial routes and scales. In her work on astronomical observatories in Hawaii, Hi'ilei Julia Hobart shows how scale-making technologies like telescopes simultaneously occupy Indigenous spaces and render outer space as a continuation of occupiable space. Her analysis calls attention to the tactics of scientific scaling versus otherwise senses of near and far.[19]

Even as vertical scale models become problematized, many ethnographers, including us, continue to rely on vertical scalarity to scaffold our inquiries and to communicate a shared sense of how phenomena connect. We acknowledge that normative social scientific uses of proportional and linear scalarity still have discursive traction. However, even when researchers are clear that such low-to-high or small-to-large orderings are socially produced, they often continue to track their project elements up and down the multiple nodes of linear or hierarchical scales. These scalar lines emanate from the here-there or this-that binaries that are common in ethnographic research and communication.

Let's circle back to an example of vertical scaling mentioned in the beginning of this section: "local/global." Ethnographic project designs often use vertical scaling to organize plans for data collection around a progressive binary. The elements of a vertical local/global scale look something like: the home → *the local* → the urban or rural → the regional → the national → the transnational → *the global*. Embedded in that scale are terms like *home* that take a predetermined spot on a seemingly normal/natural schema of small-to-large or here-to-there forms of living and belonging.

As an example, imagine a project focusing on unhoused people in San Francisco, California, a West Coast city in the United States. The researcher might take a standard approach to defining the scope of the project by placing the *unhoused individual* into a local/global vertical scale. In this scale, *home* is put into a conceptually nested or descending situated order of empirical and theoretical sites. The scholar might follow this scalar order by researching how a person becomes houseless, how houseless people utilize shelters or other dwelling places, how they search for jobs and housing, how they find joy and care in their lives, and then how Western houselessness can be contextualized in terms of national and global territory. This is a vertical processural understanding of how a single ethnographic research object ("the houseless individual") moves up or down through progressive spatial scalar nodes, that is, the local home to national territory. We represent this vertical schema and its data-collection planning in Table I.1.

TABLE I.1 Vertical Scaling

Vertical Scaling Framework *Follow the Being or Thing Up or Down*	Project Elements in Relation	Vertically Oriented Data Collection Questions
National or global	Unhoused individuals *in relation to* the nation	How are the unhoused in San Francisco situated in relation to national forms of belonging?
↕	Unhoused individuals *in relation to* private property regimes	How are the unhoused in San Francisco situated in relation to private property regimes?
↕	Unhoused individuals *within* social coping processes	How do the unhoused in San Francisco respond to the Western national politics of bare life?
↕	Unhoused individuals *in relation to* the workplace	How do the unhoused in San Francisco search for and hold jobs and housing?

↕	Unhoused individuals *in* shelters	How do the unhoused in San Francisco utilize shelters or other dwelling places?
Local	Individuals *in relation to* homes	How do individuals in San Francisco become houseless?

Don't get us wrong: there is nothing inherently problematic with a vertical conceptual schema like this and its thorough fieldwork aims! However, for the ultimate purpose of theory-making, vertical scaling often tracks one central social group/thing/process (e.g., houseless people) in a way that reproduces a predetermined spatial and ontological "scale of things in the world." If you adopt only this approach, you might overlook ways to study houselessness by constellating it within a less regimented linear conceptual framework. In addition, you might fall into a pattern of adhering to normative perspectives on what is central or "ex-centric" to lived experience.[20] You might find your curiosity being foreclosed.

Multidimensional design allows researchers to create a conceptual framework that assembles a variety of elements—including places, people, processes, and things—that might officially "belong" to different scalar or ordering schemas, or to no scales or orders at all, without imposing a hierarchical or centering order on the whole project.

You may still be wondering about how to hold this kind of perceptual space given how complex a project is. You may be thinking "It's impossible to cover it all! I can't do everything!" The ethnographer's way is not about covering or doing everything. No project begins and ends with "everything" in the frame—to try to do so would mean, as we have heard wise scholars such as Sean Brotherton say to students, that you never get out of the field! The ethnographer's way is about finding a manageable framework of multidimensional connections that allows you to engage the enormity, complexity, and dimensionality of the project's unbounded lifeworlds. It is about creating a project that can *actually* be done! This is important because having a diverse collection of elements isn't necessarily "too much" for a project to contain. Rather, the problem is the quality of the assemblage that allows concepts and fieldwork to hang congruently together. Therefore, when we talk about multidimensional space, we are referring to an actual construct that you create and within which you can manage, complete, and care for a successful project.

For example, a multidimensional project on houselessness in San Francisco could constellate inspired curiosity about relationships between lives-without-homes and the imperialist housing industry. Such a project would invite a framework that connects a broad but not necessarily linearly defined scope of processes and contexts. This might result in a slightly different take on, and potentially more unusual and innovative divergence from, the vertical schema above. To do so, a research project designer might place houselessness, as a constructed conceptual condition and political economic category, into promising ethnographic relations with other key elements: people and other beings, places, things, and contexts. These elements could be found in the following processes and sites, just to name a few: the fluctuating dynamics of unregulated real estate capital within transnational speculative real estate markets; racialized redlining as an enduring logic of slavery, militarization, carcerality, and colonial ordering; government-business complicities in racialized and classed gentrification; the way the production and responses to houselessness erase Indigenous as well as otherwise concepts of homespace and belonging; the relationship between houselessness and the denial of home-space to more-than-human-beings; and frozen wage labor in relation to climbing housing costs under changing capitalisms. All of these have local-to-global or small-to-big dynamics, but the research site plan and data-collection questions don't have to orient at that level.

As a result, the framework of inquiry might include some or all of the vertically oriented processes named in Table I.1 but could also add additional dimensions that *scope out* the meaningful place of houselessness within contemporary spatial and temporal structures of US property and territoriality. Table I.2 indicates some possible multidimensional project foci.

This kind of multidimensional investigation could result in original theoretical insights on race, class, territorial, and speciesist formations that structure particular kinds of chronic houselessness *and* chronic homebuilding. This kind of unexpected but powerful research connection-making is what can make ethnographic work so mind-blowing and lastingly impactful. It is what well-practiced ethnographers learn to create over time, which we present here as something available to all at early experiential stages.

Designing projects that include multidimensional conceptual connections open up new modes of curiosity and new analytical possibilities. While

TABLE I.2 Multidimensional Frameworks

Multidimensional Framework *Interconnecting Project Concepts*	Project Elements in Relation	Dimensionally Oriented Data-Collection Questions
Real estate sales and loan practices ↔ Transnational speculative real estate markets	Homes as residences *in relation to* homes as financialized forms	How do San Francisco real estate and banking practices connect to broader unregulated real estate speculative processes?
Government policies ↔ Gentrification	Government un/regulation of real estate practices *in relation to* the corporate housing industry	How do government-business complicities in San Francisco shape the gender, race, class dimensions of houselessness?
Rural dispossession and development processes ↔ Urban houselessness	Colonial territorialization in relation to "houselesssness" as a marker of unpropertied life	How do government and development processes in San Francisco connect to US historical colonial projects to manage territory as white property?
The management of more-than-human spatial homing/displacement ↔ The management of human spatial homing/ displacement	Human houselesssness *in relation to* more-than-human displacements	How do the experiences and spaces of human houselessness relate to the government and corporate management of plants and animals?
Labor spaces ↔ Home spaces	Housedness and houselessness *in relation to* labor and production	How do houseless people in San Francisco fit into or escape labor structures?

anthropologists are quite conscious of binaries and progressive hierarchies, they can be trapped in unexamined attachments to vertical-only conceptual and data-collection schemas. Such trappings can lead to an analytical method that reproduces the very thing that they want to avoid—a preordained, foreclosed model of a linear or binary-ordered world. A totalizing politics is at work in this placeholder default: the hegemonic imaginary of *the* body in *the* world or in *the* globe or on *the* planet conjures certain particular spatial and scalar progressions, as well as the ordered relations of (white) Western nationalism, that continue to be naturalized and used in discourse.

In this handbook, we provide many examples of multidimensional project design, from straightforward examples to those that are stunningly audacious. All examples refuse boundaries that prescribe what is inherently and intrinsically relatable across categorical differences.

But before we get to these examples, we offer a moment to imagine multidimensionality. To orient your curious imagination, we ground in the nonlinear, nonvertical, and nonorderly relations that may greet us in more-than-human worlds we walk through. We hope this brief meditation draws attention to a broad noncentered expanse of temporal, material, and spatial multidimensionality. Going into this imaginative zone will evoke an elementally and organically entangled, rather than a quantitative, minimum-maximum awareness.

To begin, imagine something *scaly* as a living being in a set of multidimensional relations, rather than thinking of it in a quantitatively scalar way. Think of the vibrantly twisting, pushing, climbing, digging, coiling, overlapping, variegated scales of a living snake in motion.[21] Its scales have different sizes and are arranged in chains that enhance perceptions of its snaky length, position, and movement. Most forms in social worlds are made up of elements like the snake's exterior scales: interconnected parts that call attention to something's relative shape and position vis-à-vis other parts of the form or other things. The snake interacts with these things not in any other order than what it experiences at the center of its own experience.

Now, perceive the snake in its multidimensional setting. Its life intersects with stands of trees, flows of air, roads with traffic, earth moved by weather or machines, settlement, genocide, revolution, property, extraction, community. Depending on what you are curious to understand about this snake, a snake-centered project can be designed to include, and to relate, such different things and processes with their unique dimensions. Such a design can be done without resorting only to a total scalar ordering that situates snakes in preexisting orders like body/nature, animal/human, local forest/whole nation. Nor does it require, for example, an ordering that would "scale" snakes as inherently *local* and air as inherently *global* (or, conversely, snakes ordered as globally responding to unspecifically local airs). Snakes and human communities and air relate in dimensional ways that are connected but different, and they require attention to those differences in order to tell an ethnographic story about their relationship. Snake-life is snake-in-relation, sometimes with things-in-relation that we can perceive easily and sometimes with things that we sense and that come to us when we drop preconceived notions and take it all in.

The disciplinary challenge for multidimensional ethnographic project design comes with the task of identifying a manageable set of elements-in-dimension to relate, and finding a conceptual "object"—like snake-life in motion—with which to hold everything together.

Identifying a Multidimensional Object

We have found that the most promising research projects are multidimensional in ethnographic as well as theoretical and disciplinary terms. As we have indicated so far, this means that they bring different kinds of concepts and scholarly literatures into relation with one another. In order to make this kind of project feel coherent and congruent to yourself and others, you must articulate how the project's empirical conceptual dimensions (e.g., the people, places, parts, and processes you will actively study) integrate with its broader conceptual and contextual dimensions (e.g., its disciplinarily relevant theoretical scope and its capacity to illuminate broadly extensive conditions in the world).

We have found a way for research designers to get, keep, and communicate this integration of disparate parts: articulating a short phrase we call the multidimensional object (MO). The MO extends the "object of study" in social science research into broader multidimensional territory.[22] An effective MO conveys both the cohesion and the tension among project concepts. In short, it conveys a project's *tensegrity*.[23]

Tensegrity is an idea we find useful when it comes to doing multidimensional concept work. Tensegrity is a term from architecture and art, particularly the work of R. Buckminster Fuller in architecture and Kenneth Snelson in art.[24] Recently it has provided a way to describe the integrated (fascial and other) structures of mammalian and unicellular organisms, and in these contexts it is designated as *biotensegrity*. In such forms—which include bacterial bodies, human bodies, and geodesic domes—nails, brackets, and skeletons aren't the things holding these forms together. Rather, the *forces* of interrelational tension and situated compression keep the parts of bacterial cells in touch with one another without skeletons; enable vertebrate and invertebrate bodies to stay integrated when their tendons, muscles, and bones push and pull against one another; and enable geodesic domes to stand open and unsupported by central beams.

Tensegrity results in nonhierarchical connectivity that distributes throughout elements evenly, creating a hanging-togetherness. Unlike in a vertical scaling process, which requires a hierarchical and linear structure like local/global or center/periphery to hang a project on, our process allows project concepts to stay in a harmonious- and tension-integrated dynamic. Tensegrity, therefore, is a sense-based term to remind ourselves that unusually strong ethnographic projects have parts that may not all seem to belong together in linear or positivistic terms.

In Module 6 you will creatively articulate a multidimensional object (MO) that evokes the dynamic relationship of your project's key concepts and keeps them together in a cogent, congruent assemblage. The job of the MO is to make perceptible the connections between project elements, from the literatures to the project description to the research questions. Unlike an unwieldy project description, the short and sweet MO is *wieldy*; it helps you handle project complexity and breakdown. It's flexible and can be revised as you iterate your project concepts and elements. But more than having design functionality, it's also a vibrant phrase that can inspire you and others about your project. (We hasten to reassure you that your project's tensegrity will withstand bouts of reorganization without completely going back to the drawing board!)

Here are the stories of how several MOs emerged in design courses we taught. In this handbook, we include student project examples, with permission, that we render to illustrate processes and exercises. We obtained permission from these researchers to refer to their projects, and they reviewed and approved the examples we rendered. Those projects continue to change as the researchers produce ongoing data and write their ethnographies. We are extremely grateful to them for partnering with us to feature their projects. Throughout the handbook, we expand on these and other examples to orient you to multidimensionality and the MO.

Tariq Rahman, whom we described in the prelude, wanted to design an ethnography of contemporary Pakistani social life. His project took into account an experience-based intuition about two distinct but intersecting processes: real estate trading and global processes like transnational financial networks and imperialist militarism. During his fieldwork, he intuited that there was a vital connection to understand between new ways and reasons for buying and selling plots of land in Pakistan (dimensions: economy and territory), financialization technologies (dimensions: technology and actuarial practices), and US imperialist policies implicating the nation (dimensions: law, politics, and militarization). Typically, he would have to make a choice to center one or two dimensions and make the others secondary. To design a project that holds such seemingly disparate connections in equal tension with one another is challenging or might be deemed a "stretch." Yet a multidimensional project design can hold together such processes.

Rahman's concept work allowed him to design a plan to investigate the interrelation of financialization and the war on terror, even if that relationship was happening at a geopolitical distance from the urban plot market in Lahore. To do so, he needed to add more theoretical dimensionality to what a "real estate plot" is in the worlds he works in. He also had to understand the relational properties of plots that connect plots to geopolitical processes. In doing exercises about this, he achieved a conceptual breakthrough. He found that several emerging events, dynamics, and technologies drive plots to be traded less like concrete forms of *land* and more like fluid forms of *market stock*. This allowed him to identify the project's multiple dimensions: land speculation as halal in Islamic practice in Pakistan (versus other forms of investment that are forbidden), the war on terror driving specific speculative practices, digital media used by diasporic Pakistani investors to drive market values, and government officials who digitize landownership while holding on to the long-term memory of land among generations.

At this point, Rahman had the beginnings of a succinct research description (RD), which we define as a short narration of what the project is about and its potential significance. In our process, the RD (Module 5) is not the end of the congruence-making process: it's the beginning. The next step is to identify an MO. As Rahman completed several iterations of an MO, he refined his original dimensional connection between land and stock.

Using "liquid land" as his MO, Rahman's multidimensional design allowed him to do a project based on a radically different investigation of financial speculation, whose forms and analysis are often limited to the generalities of

North American, European, and East Asian markets. He was able to retheorize "land" and "liquidity" as financialized *processes* emerging in everyday structural and cultural settings as well as in relation to hegemonic and imperialistic processes of labeling places as sources or targets of (counter)terrorism.

Multidimensional objects can take various forms and can help multidimensional projects, whether they take place in single or multiple fieldsites. Remember Forest Haven's project that brought the environmental and sensory dimensions of food politics together? Her MO was "food sensing." We come back to her project several times in the modules ahead to illustrate how this MO helped her design a unique project on Native food sovereignty. Let's look at two additional MO examples.

Jason Palmer, another researcher, designed a study of Mormon conversion that prompts pilgrimage or migration from a sacred site in Peru (as noted in the Book of Mormon) to Mormonism's Zion, Salt Lake City, Utah. More broadly, he focused on how Peruvian Mormons navigate trans-American ties through the production of sacred places, where evangelization based on heavenly notions of borderless relationality encounters the obstacle of the sovereign US border. By doing concept-work exercises that enabled him to creatively conjoin the legal, economic, and spiritual stakes of the project on equal footing rather than in a hierarchical way (e.g., assuming the economic is more salient than spiritual), Palmer devised this MO: "im/mobile memberships." Here the contingent mobilities of pilgrimage and migration relate *across processes* within family, congregation, state, and earth/heaven. He could ask questions and look at experiences and objects that didn't force a choice between attending to either national or religious forms of belonging.

Holding all these differently scalar elements of the project together via "im/mobile memberships" allowed Palmer to relate two seemingly disparate dimensions of life: the Mormon religious definition of and care for its members' souls, and processes of national membership and border maintenance. By bringing these dimensions together, he could do two things. He could design a project that attended to the laboring, material, and traveling person moving up or down an economic vertical scale. But he could also go further. Palmer's MO, with its attention to the embodied and soul "member," allowed connections between multiple and nonintuitive project dimensions: religious experience, economy, migration, hemispheric geopolitics. That, in turn, opened up imaginative thinking about the soul as connected to encounters with the state. Palmer's multidimensional project will contribute to retheorizations of religious experience that are informed and directly connected to these other domains, thus retheorizing the dimensionality of those domains as well.

Another researcher, Kimberley D. McKinson, set out to study how violent crime is represented and instantiated in Kingston, Jamaica. Instead of rendering a vertically aligned urban anthropological project that centered *criminal* or *victim* along the pathway of crime → court → state → nation → geopolitics, McKinson did something very different. She followed her curious hunch that contemporary forms of security involved different spatial and temporal dimensions of experience: the historical legacies of control in slavery and plantation life, the relations of people in systems, and the uses and reuses of everyday building materials. This allowed her to include in her project her intuitive interest in the metallurgical and architectural designs (like gates and fences) of securitization in Kingston residences. Her MO, then, was "security ecologies," a framing that allowed her to draw together the multiple dimensions of her study: the historical disciplinary use of metal during colonial slavery; the aesthetic and recycling use of colonial and contemporary materials in residential fortification; the varied use of surveillance and security technologies to avert crime; community practices around ideas of safety, such as neighborhood watches; and the use of slavery's security technologies such as patrols, passes, and curfews.

With the MO "security ecologies" holding these ethnographic and historical project elements together, McKinson was able to pose unusual questions about historically and relationally racialized subjects (that discipline and are disciplined by security technologies and practices)—ones who performatively reengage the territorial and material conditions of colonial slavery in the context of discourses about "security." If the MO had been "landscapes of criminality," it would have directed her to conduct an ethnography of crime, a standard colonial and racialized concept, in relation to land and property. But because she attached the word *ecology* to *security* she created an effective multidimensionality in her project that enabled her to craft a genealogy of the Black body's relationship to violence and material transformations over time in Jamaica, from slavery to the contemporary postcolonial moment.

It is important to note that even though these three examples happen prior to fieldwork, they emerge from deep engagement with the field and they often change over time.

We have found that both students and colleagues consistently experience the "truth" of a congruent MO in their own bodies. There is an experience of elation, joy, and sudden lightness that resonates powerfully—much like an apprentice begins to get a feel for effortlessly and intuitively enacting her craft. This deep knowing emerges after a lot of trial and error of connecting project elements together and testing for congruence across dimensions and then finding that it just doesn't seem right yet. Moreover, researchers doing concept work together find a literal resonance among all bodies in the group as they arrive at project congruence. Because our orthodox vertical foundations for building projects are not there, the body stands in as the tuning fork that intuits and resonates with what feels right.

Another aspect of research design and felt tensegrity that a multidimensional design process clearly addresses is that of how and why to articulate questions.

Relating Multidimensional Zones of Inquiry

Our handbook brings you to the research question–asking stage midway through the design process, rather than forcing questions to emerge before you have the strongly felt sense of a project's elements, processes, contexts, and concepts that the MO represents. Often research designers don't acknowledge that there are different kinds of "research questions," that they have different forms and functions, and that those differences need to be clearly distinguished. In Modules 7, 8, and 9, we help articulate *three* different kinds of research questions necessary to create a well-defined and congruent framework of inquiry. We make the question-type distinctions clear and highlight the project design pitfalls in not understanding them. These question-type distinctions are embedded in different *zones of inquiry*—what we refer to as the scoping, connecting, and interacting zones.

In the scoping zone, designers will develop an *overarching scoping question* for their projects; in the connecting zone, they will formulate *data-gathering questions* that pertain to project-clustered concepts; and in the interacting zone, pertinent *field-based questions* (interviews, archival, etc.) will be generated. Multidimensional research emerges when each inquiry zone is consistently and ethically put in relation with the other. The scoping zone defines the overarching theoretical and social field, as well as the project's big question and overall significance. The connecting zone opens all the interrelated possibilities (obvious and intuitive) of data collection in relation to all the project elements. The interacting zone focuses on the intellectual and social aspects of fieldwork, allowing designers to plan modes of purposeful and ethical inquiry through observations, interviews, and participation. We come to methods last in this handbook for a reason: methods are meant to be in service to the project conceptualization process and not as the determiners of it.

As the conceptual and ethnographic element that links these zones, the MO helps you go back and forth among these and other module elements as you enhance your project's aims and congruence.

Project Listening: Attending to What's There and What's Possible

By now you may be imagining the expansive possibilities of multidimensional research design. In general, projects must emerge and expand in order to get their scope. But the expansion process can become unwieldy in every phase: design, fieldwork, archival research, postfield analysis, writeup. To manage productive expansion and contraction, you need to continually iterate and refine your research description, MO, research questions, and data-collection foci. This helps you to maintain an ongoing feel for the project and helps you navigate your intuitive and emotional relationship with the necessary upheavals and transformative moments of project design. We provide ways throughout this process to help you stay aware of the project's internal congruence and its external relationships with other projects and processes. Project listening is the process of attending to these dynamics in your own project and those of others.

We like "listening" because it expressly pushes against—but also adds to—the "speaking" or "writing" processes of expression. Project listening is not about hearing in a biological sense; it is about adopting a stance of curiously perceptive stillness and intentional spaciousness. Stillness allows unwavering attention without distraction, and it can help free the mind from immediately analyzing its encounter with some aspect of the project. When you are feeling stuck or unsure about your project, stillness enables you to have a direct, embodied experience with the elements, questions, ideas, and connections that animate your work. It also stops you from defaulting to abstract or analytical "solutions" that are not grounded in the project elements themselves. Spaciousness offers literal and metaphorical room to widen our views; it helps us learn when to pause, stop, push forward, and hold on to our concept work, leading to openness and receptivity. Taken together, stillness and spaciousness are key to cultivating wisdom (not just knowledge) about our research.

In this handbook, we provide many project listening exercises that help you to slow down, reflect, and assess in order to reconnect with the project's interconnected dynamics. For example, after you identify your literatures and project concepts in the first three modules, Module 4 allows you to revisit that process in relation to key previous developments, namely your literature selection. Iterating these key design features creates a productive tension that allows the project to expand when it can and contract when it must (thus, again, this is not about bringing "everything" in). This maintains the project's congruence and its possibility for significance.

How does project listening help you feel and support project tensegrity? Project listening makes it possible to "stay present with the project" throughout its life stages: open-ended imagining, curiosity-based designing, conducting research, analyzing, and writing up. It helps researchers stay in touch with the relationship between personal and broader stakes and intentions. Staying present with the project is therefore also a way to stay present in community. With others, we develop skills to be present with the project as it emerges: paying attention, practicing conceptual unfolding, untangling layers, and allowing insights to shape new directions.[25]

Project listening attends to the lifeworld of ethnographic research, making space to open centuries of subjugated knowledge while valuing our collaborators in knowledge and theory-making. If the creative works that we make together in this mode can be thought of as aspects of living consciousness, then we are staking out a path to wake up to necessary ways of perceiving and understanding the multitudes of worlds.

Overview of the Multidimensional Design Elements and Processes

This handbook is designed to help you begin a project or rework one that you've been dealing with for years. Each of the ten modules contains a short explanation of the project element you will design and its purpose; exercises that help you attain those outcomes (with examples provided); and prompts for how to process your work in groups. We guarantee exciting and unique breakthroughs. At the end of each module, you will summarize your project elements in a concise project grid. You will have a fully fleshed out, but cogent, view of your project that will be immensely helpful in orienting you toward writing a project proposal, thesis proposal, grant proposal, dissertation, or book. Table I.3 provides an overview of each module's project elements and processes.

TABLE I.3 Handbook Overview

Symbol	Section	Project Element	Process and Purpose
	Interlude 1	Community Work Plan	Consider your and your groups' needs; learn to set up a community feedback space.
	Module 1 Imagining Research	Research Imaginary	Narrate your project's research processes and possibilities; and create space for known, intuitive, and unknown project concepts.
	Module 2 Focus on Literatures	Literature Relationships	Link project concepts found in the research imaginary to three project-defining literatures.
	Module 3 Map Concepts	Concept Map	Create a one-page map of your project's key concepts.
	Module 4 Create Multi-dimensional Concept Combos	Concept Combos	Locate a key concept combo that represents your project's multidimensional contribution to literatures.
	Module 5 Describe Your Research	Research Description	Develop a succinct research description that includes broad key concepts; clarify your project's core multidimensional form and possibilities.

✳	**Module 6** Perceive Your Multidimensional Object	**Multidimensional Object**	Craft a phrase that evokes the project's conceptual multidimensionality and structural tensegrity.
	Interlude 2	**Inquiry Zones**	Learn about project inquiry zones and how to integrate them.
⌒	**Module 7** The Scoping Zone	**Scoping Question**	Scope out the project's overarching research question and significance.
ɰ	**Module 8** The Connecting Zone	**Data-Gathering Questions**	Form data sets and data-gathering questions that will provide the data needed to answer and theorize the scoping question.
>–<	**Module 9** The Interacting Zone	**Interacting Questions and Methods Plan**	Generate field-based questions that connect directly to methods and data-gathering questions; transition to the communicating and planning phases of project work.
(symbol)	**Module 10** Mobilize Your Research Project Grid	**Gridwork**	Use your project grid to prepare ethics boards applications, write grant proposals, and do multidimensioning during field research.

We suggest that you do the modules in order. The exercises in each module build on the previous in an iterative manner and also involve doing reflections on previous work, which we show you how to do in Module 1. These intentional processes help you to develop strong intuitive mastery of project design—the ultimate goal of this book—and not just a great research project.

As you move through the modules, you'll find that we use the term *multidimensional* as a noun, an adjective, and even a verb. Sometimes we want to remind you that you are not just making a multidimensional project, but you are also actively multidimensioning literatures, project elements, and other project features. Your multidimensional design is an action and way of positioning yourself in a creative space as much as it is a concept-work product!

The connective tissue of this design process—concept work, imaginative and intuitive curiosity, multidimensioning, project listening, and working in community—conveys the enlivened spirit that animates *The Ethnographer's Way*. Know that the liberatory possibilities for research, and life that you wish to live, can be imaginable and shareable.

Create a Collective Concept Workspace

We have found that a project reaches its best imaginative manifestation when researchers do iterative concept work with other researchers—especially at the raw beginnings of project conceptualization. This certainly applies to disciplinary peers, such as anthropologists working with other social scientists, or to collaborators working on a single project together. But it especially applies to researchers doing individual projects who can then practice concept work together. Whatever the configuration, we refer to this activity as *collective concept work* that takes place among a *community of researchers* who intentionally come together to form *collective concept workspaces*—the physical or virtual dialogical spaces that are dedicated to project crystallization.

These work-process suggestions are best instituted in tandem with an exercise submission plan, as described in the prelude, for choosing which work products your collective will be reading and responding to in each module.

INTERLUDE 1 PURPOSE
Consider your and your group's needs; learn to set up a community feedback space.

A collective concept workspace is an interpersonally conducive and pro-embodiment space that helps designers discover project elements and clarify social commitments together. By foregrounding the collective and intuitive aspects of concept work, this approach shifts research work out of cultures of separation and structural violence and into shared modes of curiosity, connection, and compassion. This shift is not easy and is impeded by racial, heteropatriarchal, ableist, classed, and colonial implications that are embedded in bodies and institutions. But we have found that attending collectively to the earliest conceptualization of a research project helps address dualistic and hierarchical epistemologies and ways of being in research institutions. That is, collective concept work fosters senses of accompaniment rather than of aloneness. It can counter modernity's racialized, gendered, and competitive logics that harm bodies and minds. It mitigates disheartening experiences of disconnection that occur throughout institutionalized landscapes, such as those between professors and students, people working in and outside of bounded organizations, and those with expertise versus those with experience. Doing consistent, collective concept work means that peers are not just treated as colleagues or as potential consumers of final products but as codesigners of living projects. Here we are choosing *not* to participate in modernity's intellectual work modes that have long been structured around the lone, isolated individual, even if we are choosing to take part in the institutions that house those modes.

To counter the damage such work patterns have created, we begin by assuming that *all researchers possess an inherent inner knowing that can be enhanced through supportive interactions.* The aim is to shift out of needing each other for competitive comparison (which invigorates systems of oppression) and instead lean into needing each other for interdependence (which energizes potentials for collective liberation). If we can see that these conceptual endeavors are always *for* collectives, it becomes possible to imagine living different anthropologically attuned lives, inside and outside of institutions. That is, we see a collective approach to research design as having the potential to shift consciousness around what it means to do intellectual work.

EXERCISE 0A Prepare for Design Work

1.   *Set up your personal space.* Before beginning your concept work and creating concept workspaces, we suggest you find a personal workspace that is conducive to your needs. If you are working

in a shared workspace where quiet attention is needed (and difficult to come by), spend some time thinking about how you plan both your space and time. This could mean that part or all of your work happens away from home, perhaps in an office or coffeehouse. Whatever the case, treat your workplace as a sanctuary by first deciding on the kind of atmosphere you need to support reflection and concentration. Consider adding images, plants, and/or objects that bring a sense of ease and support. Have a special place for water, coffee, or tea. Add a pillow or other props to the chair to support your body as you work, and have other self-care supplies on hand. Think about the time you need to walk and stretch. The purpose is that you find a state of being that nurtures your capacity to create, reflect, rest, listen, and do processes again and again as necessary. Some people find it helpful to do a ritual to create space, external or internal. Or to begin work with a moment of silence, a prayer, or a formal meditation to honor the shift from life's various activities into the space of creative concept work. This is your space—and within it is the spaciousness you need to create this project.

Be generous and kind to yourself in discerning your needs, as these small things go a long way in caring for your body, mind, and long-term concept work. When setting up collective space inside or outside of a classroom, you can do the same or similar process for the group.

2. *Gather concept-work materials.* After you have made decisions about your supportive space, you will need to gather some materials and decide how and where you will store all your work. For work by hand, we suggest these items:

   a. Two blank notebooks, one that is lined (or not) for writing and one that is blank for drawing, as in an artist's sketchbook
   b. Pens and pencils, including highlighter pens
   c. Sticky notes
   d. File folders
   e. Whiteboards

For digital work, document archiving, and project design research, we suggest these items:

a. A word-processing program that is compatible with your group members' preferences and tech resources

b. A designated folder on a stand-alone or shared drive (for personal and group work) or a file organization program such as OneNote

c. Access to the internet to search for scholarly and creative works and for project-support software tools (as suggested in the modules)

Collective Concept Workspace Possibilities

The exercises in this handbook can be done individually. However, they are composed to address designers working collectively in order to facilitate supportive and insightful feedback. In a classroom setting, most concept workspaces take place during small-group dialogues and so the workspace simply comprises the members of your class. Beyond the classroom, choosing members to work with usually involves personal and collective criteria.

For those starting out who do not already have a collective work tradition, we offer ideas on concept workspace size and processes: a working group of three is the magic number (although two is better than one, and four is better than five). However, you and your companions can innovate meeting configurations by creating a large group that contains smaller groups for interactional feedback. Smaller group configurations can be permanent (with the same membership) throughout the duration of the project/modules. Or they can periodically rotate members from the larger group into smaller groups at agreed-on intervals. We have found that an occasional (not frequent) rotation can offer fresh questions and insights that were not previously attended to within one's permanent group of three. If you choose to try this, then consider doing the rotation at Modules 5 and 8. When permanent groups reconvene at Modules 6 and 9, members should read two modules of work because both Modules 5 and 6, as well as Modules 8 and 9, are interrelated.

Periodically, small groups can come together as one large group to check in and talk through design questions and other issues. In formal teaching or workshop scenarios that unfold over days or months, we usually open each meeting with a large-group individual check-in/discussion, focused on the contents of a handbook module, before breaking into smaller working groups to discuss weekly module work. Check-ins help designers express their delight or their confusion that others in the larger group may be experiencing. These shared reflections provide support for students. They also provide the opportunity for instructors to address easy-to-clear-up confusion as well as reassurance about the process when students express dismay or frustration. Doing this at the beginning of a class meeting can refresh the atmosphere and reset negative or disheartening student stances before they start their group work. When this happens, breakthroughs in the feedback process become far more possible.

These are just a few examples of what's possible for sizing and maintaining a collective concept workspace. We encourage you to assemble what's best for your own needs and objectives.

The Importance of Agreements

Before you engage in concept workspaces, we suggest that you establish agreements for building community to work on the modules together. Agreements are essentially about honoring and holding space for each other's personal needs and boundaries as well as for the collective dynamic. But they also provide guideposts for avenues leading to trusted engagement. As Prentis Hemphill and adrienne marie brown have pointed out, boundaries are not just about putting up walls or creating protection, but about accessing connection by "making room for the important stuff to come through—[and] trust[ing] that this important stuff has its own force that is putting things in proper perspective."[1] Accessing connection by "making room for the important stuff to come through" is, for us, the essence of an embodied collective approach to research design.

If you are not accustomed to knowing and actively working with your boundaries, you are not alone. As brown and Hemphill point out, "Trying to find a sweet spot with boundaries can be a challenging calibration process.... Boundaries are not a destination to arrive at as they may be shifting as you grow into the process." Boundaries not only open a generative space for engagement, but also help reverse conflict avoidance. They also provide the groundwork to *lean into* disappointment, conflict, breakdown, or other difficulties when doing collaborative work. This process supports each other's full autonomy to engage challenging emotions. That is, agreements widen the view of knowing when to act, pause, sit with, slow down, support, or let things be. They increase the chances for authentic and ethical engagement.[2]

This may sound too woo-woo for academic environments. But we all know what happens when there exists no conscious framework that holds agreements. Interpersonal challenges can arise and get ignored, and the dynamic among colleagues becomes increasingly grim until everything disassembles. Or trouble gets expressed in unskilled, harming, or covert ways, which encourages increased disconnection and unworkability among group members. It's a mode of processing that's not interested in resolution; rather, it gets invested in rendering problems unresolved while taking up years of time. When no conscious agreements are in place, then agreements held by systematic oppression dominate and fill the institutional vacuum: power tripping, bullying, belittling, sabotage, exclusion, sexual and racial harassment, and other traumatic experiences that remain unresolved through academic lifetimes.

When we don't have boundaries or agreements, then we let the violent expression of our institutions dictate how we relate to each other.

So, when we talk about the potential for a liberatory way of being in our intellectual worlds, we mean daily counterinstitutional practices that provide possibilities for intellectual transformation. The simple act of consciously creating agreements has the potential to rewrite and course-correct personal and institutional futures.

Agreements should include your vision for working together and how you bring out the best in each other's work as well as manage participation challenges. Exercises 0B and 0C help you to lay the groundwork for compassionate and durable working relationships.

EXERCISE 0B **Personal Needs Assessment**
for Collective Concept Workspaces

Building community begins with considering individual needs and preferences *before* the first group meeting. This exercise is a simple "personal needs assessment" questionnaire that can aid this process. We encourage members to answer these questions in writing (not for sharing, but rather for personal brainstorming) in order to clarify boundaries, needs, and willingness to participate in community. This helps establish what's necessary for members of this particular group in order to build a mutually conducive workspace. Your answers provide the raw material for negotiating group agreements:

a. What do you like about working in groups and with others?

b. What don't you like or fear about working in groups and with others?

c. What are the most important things that you need working in a group?

d. What do you not want from group members?

e. If any other questions or concerns come up for you, explore those in writing as well.

EXERCISE 0C **Make Collective Agreements**

This exercise takes place among your group and constitutes your first meeting. You will set up and establish a grounded collective ready to embark on research design concept work (which could possibly take longer than one meeting). We offer items here that can guide the coproduction of meeting agendas. Here are possible agenda items, in the order that we find most conducive:

1. *Introductions.* Even if you know each other, talk about yourself, your research, and how you arrived at your particular project. What do you like about your project and why do you specifically want to do it? Give some time for everyone to really talk through this.

2. *Community agreements.* Everyone should draw on their own personal needs assessment to contribute to a conversation that will lead to the group's guiding agreements. Effective working groups begin by coming to a set of agreements that lay out how the group will connect and interact. This is especially important within multicultural, multinational, multigendered, and multiclassed groups. Community agreements include how you will interact, manage challenges, and deal with responsibilities and logistics. When established, they cultivate group

members' well-being while diffusing competitive interactions, and in so doing they help to open an otherwise way of being in academic, or other, institutional life. Again, take time for this process. Know and openly acknowledge that determining these agreements might require you to stretch your vulnerability and be willing to openly share the needs and boundaries you identified in your personal assessment. You can talk about these topics without oversharing about your fears or experiences. Being clear and open with each other creates the basis for strong agreements that hold the community together.

Know that down the road community difficulty, challenges, and breakdown are possible. This usually happens because needs and boundaries either were not discussed or were unknown or unclear to begin with. There is nothing wrong with this; when breakdown happens, the group can lay out their options in terms of going forward. Consider talking through the challenges and breakdown and how you will manage participation challenges. Consider revising your agreements to accommodate more clarified individual and group needs. The following example list of agreements is derived from two different graduate courses we both taught.[3] Each group can decide on agreements that are relevant to their own practices.

We agree to:

–   Create and practice holding space for compassionate accountability. This is the ground for addressing conflict or tension that may arise doing group work. Recognize that we are all learning and trying.

–   Strive to make group work an act of compassion so that projects and research designers get to stretch themselves into their greatest possible iterations.

–   Practice using compassionate language that helps to push each other further.

–   Get to know how each small-group member prefers engagement and feedback on their projects. Practice bringing care to each other's boundaries.

- Communicate with each other about what feels right in terms of written and oral feedback.

- Consider positive questioning in giving feedback. Refrain from "shoulds" or giving advice unless solicited.

- Strive for mindful listening, which allows for everyone to be fully engaged in what each has to say. Allow silence to be an active part of group engagement.

- Strive for mindful speaking. If you tend to speak a lot, practice holding back; if you tend not to speak much, practice stepping up. Recognize the context of interruptions: when they are called for and when they feel inappropriate.

- Be conscious of equitable time for everyone. Use a timer to ensure that everyone gets a turn.

- Be willing to be uncomfortable and not know the answers; trust that the answers will come in time. This is key to riding the ups and downs of concept work.

- When conflicts arise, halt the concept workspace and give the problem attention and time. This means that we might need a facilitator. Without a facilitator, we agree to take turns speaking (perhaps with a talking stick that marks who is currently speaking) while all others listen without responding or interrupting.

- Use "I" statements and speak from the heart. We agree to see everyone as fully human with struggles and needs. We agree to be as patient and compassionate with this process as possible.

3. *Feedback*. Based on the community agreements, establish a format for feedback processes. This creates a container that holds the community together at each meeting. Know that the format may need to be adjusted a few times to suit the group's needs. Especially as you get used to working together, allow room at your meetings to check in to see if the agreements or format need to be adjusted. Consider how each person gets a turn to discuss the material they produced as well as how group members offer

feedback. Always try to lean into providing support so that each project can see its fullest manifestation.

Here is a possible format for giving feedback in a non-judgmental yet powerful way. This approach is drawn from Terry Wolverton's *Writers at Work* creative writing classes in Los Angeles, California.[4] Wolverton adapted Liz Lerman's Critical Response Process, as have others, such as Felicia Rose Chavez, with her guidance for antiracist writing workshops.[5] Unlike standard academic feedback, it steers away from telling a concept worker what to do. Too much advice is like too many cooks in the kitchen. It leads to overwhelm and can block helpful ways to take concept work through future iterations.

Consider starting small-group work with a check-in (even if you already did that in a larger group) to get connected to each other's well-being. Also, consider revisiting the agreements before you start feedback.

Feedback can proceed as follows:

a.   *Meanings.* As someone providing feedback to a writer about their concept work, consider leading with the *general meanings* that you took away. For example, this could take the form of a summary, of major themes, of nonexplicit nuance, or of implied or hidden meaning. When a feedback giver does this, the writer can have a powerful sense of how the work is understood by a reading audience. This provides the ground for clearly articulated future iterations.

b.   *Noticing.* General meanings can then transition into more specific text details that stand out to a reader. Avoid leading with "You should…" and instead say, "I notice…" "I notice that I found this paragraph to be really clear and insightful." "I notice that I didn't understand the part about x." "I notice that what you write on page x was really powerful for me." Again, this kind of feedback provides strong insight for the writer who can detect whether their meaning is clearly conveyed or not. It can also clarify areas in the concept work where there are conceptual congruence, confusion, or unknowns.

c.   *Questions from the feedback givers.* The feedback givers' questions can signal what needs to be refined or clarified. "Can you say more about the relationship between x and y? What is the connection that you're imagining at this point?" The question that comes closest to giving advice leads with "What if...?" "What if you considered examining x as part of this imaginary?" Leading with "What if?" creates possibilities rather than telling a writer what to do.

d.   *Questions from the writer to the rest of the group.* Any question is possible, such as "Did it make sense to you when I wrote x?" They can also be based on the feedback just given: "I was really interested in what you said about y. What was your thinking about that? Can you say more?" A writer can ask anything they want, including "Did you like it?" Here, advice and opinion are given only when solicited.

You can use feedback like this, combine it with the prompts provided at the end of each module, or adapt your own to suit the needs of your group.

Your group should also discuss how you plan to receive feedback. In the example above, the concept worker listens to the feedback until the end, when it then becomes their turn to pose questions to the group about their work. You might find that other approaches work for your group, but we suggest listening to feedback from others without responding right away. When listening to feedback, sit for a moment and ask how it feels. The purpose here is to try to avoid defensive modes and instead lean into what feels right.

Your approaches to giving and receiving feedback will need fine-tuning as you progress. In your initial meetings, the key thing is to discuss your expectations for feedback and agree on an approach to start with.

One last bit of encouragement: stay with the trouble! Attending to the processes of mutual support in research design can require many adjustments to the way you share work and respond to challenging feedback. Staying with the trouble may require that you lean in and get more engaged, or that you pull back and take a break. Sometimes it means grounding yourself into *process*, which can be freeing, rather than grasping for, or obsessing about, conclusions, results-now! attitudes, and finalized products, which all can be anxiety-inducing.

This collective process has its own framework of inquiry. In other words, the question at hand is: *What does it mean to cultivate trustworthy, collaborative relationships and how can this be done to best meet people's needs?* Such relationships don't just happen: they are cultivated and come from discovering new ways to respond generatively to each other. The concept-work process begins and is sustained by imagining, which creates an unusually satisfying and illuminating opening to project design.

Imagine the Research

In this module, you lay the groundwork for multidimensional concept work by writing a research imaginary. A research imaginary is a freely written narrative about a project's contexts, aims, and potentials. We first experimented with research imaginaries in graduate school while working with George Marcus at Rice University. He assigned one every week, pushing us to write in detail about what we actually knew and didn't know about our intended research projects. This got ideas and hunches out of our heads and onto paper. As an informal writing exercise, the process emphasizes articulating intentions and possibility over planning and certainty.

MODULE PURPOSE
Narrate your project's research processes and possibilities; create space for known, intuitive, and unknown project concepts.

This invitation to start by imagining may appear too open and aimless, but it is, instead, a very focused process. It makes you write about the most concretely tangible as well as the most abstractly theoretical aspects of your project—from beings to social processes to broad contexts to scholarly ideas. You're already explicitly or implicitly thinking about these features of your project, so this allows you to ponder what's possible to do and express what is meaningful to inquire about, rather than what "should" be done. The imaginary is thus a form of *prospective reflection*.

By prompting you to prospectively imagine your study and then reflect on those imaginings, we flip a time-honored scholarly genre on its head: the *retrospective reflection*. Social scientists have been writing postfield memoirs for over a century. These can appear as works separate from the ethnographies themselves or as ethnographic prologues or afterwords. They bring into public view, sometimes for the first time, the author's deeply personal reasons for doing the work, the problems they had doing it, the ways it changed because of unexpected circumstances and failures, and the poignantly fraught social dimensions of interpersonal relations and data gathering.

Notable ethnographers have produced memoirs that do not maintain artificial separations between scholarly analysis and personal experience. They also often critique those artificialities. Zora Neale Hurston's personal commitment to writing what wants to be written led her to intertwine social science, autoethnography, and story-writing in her autobiography. Hortense Powdermaker's autobiography illustrates how methods must be seen as both ways of researching and ways of living and being social. Ruth Behar insists that observation is inherently subjective, which makes vulnerability a posture of ethnographic inquiry rather than a liability. Marilyn Strathern's work on gender in Melanesia relays her experience digging up layers of unexamined Western theoretical assumptions in order to really see her research topics in terms of her interlocutors' experiences.[1] The research imaginaries in this book follow these examples by inviting you to write from the get-go about research as an act of questioning, not-knowing, relating, hoping, sharing, and making worlds.

Your first research imaginary will be 1,000–1,500 words (approximately two to three pages, single spaced). You can think of it as an invigorating brain dump. This process is what brings your project concepts into thrillingly messy view. The goal of this module is to bring forth, out of this vibrant and wonderful jumble, a manageable group of *key concepts* to work with. At the end of the module, you will enter your key concepts into a project grid of design elements that you will add to, module by module. Iterating these key concepts from now forward will shape and reshape the subsequent design elements: the key literatures, concept map, research description, and research questions and methods. This module initiates the process of expansion and contraction, the creative rhythm of opening up, refining, adding, letting go, and iterating.

As the initial step into the ethnographer's way, the imaginary we coach you to write is much more than an exploration of what you'll do in the field; it can articulate your heart and spirit. As an open but focused think piece, a research imaginary allows for expansive thinking within flexible parameters. This can reveal what you are focusing on, and it can also surface hidden gaps, confusions, and unexamined assumptions. An imaginary can also be the place where you begin to materialize thoughts and feelings about why you're doing the work, about transforming ethnographic practices, and about supporting otherwise ways of being to prosper.

Such imagined visions for anthropological work and for lifeworlds are central, not peripheral, to the process of multidimensional research design as an ethical work of innovative assemblage and expression. Now that you understand the purposes of a research imaginary, perhaps you can see how it serves as a marked counterpart to the formal, objectivized grant proposal–writing genre.

Lastly, we want to note that writing a research imaginary is also not just for the beginning of a project. You can do lengthy or short reflective research imaginaries during the project conceptualization process, in the field, during data analysis, or while writing the final project. They are especially helpful when you hit a wall, get befuddled, or struggle to establish realistic limits for your work or analysis. Research imaginaries slow you down enough to pay attention to what you are about to do or what you can do. Via the simple task of narrating, the imaginary pulls you through the threshold of what's possible and into what you can and will actually do.

Do not rush this inaugural attempt to write about your project. It deserves time and attention (as do you). The writing exercises here deliberately build in spaciousness so that creativity and imagination can take the lead, rather than being an afterthought. Along the way, auditing prompts help you see how you might be more specific and intuitive about what you want to know and why, and how to let go of language that doesn't honor the complexity of your intentions. The exercises to find your key concepts will also help you balance your project's empirical/theoretical conceptual landscape in anticipation of future concept mapping. We recommend that you do the module's exercises individually and then meet as a group to discuss them using the feedback processes that you have agreed on as well as the collective concept workspace prompts at the end of the module.

The prompts for this module's first exercise unfold over four days, and they build in times for rest and reflection. However, please note that you will need *a week of sequential days with protected time* to complete the entire module.

Exercises

EXERCISE 1A Write Your Research Imaginary in Four Days

The key to writing a successful research imaginary is to adopt a free and breezy yet realistic and practical attitude. It is important to stick to clear and casual language, particularly terms that are relevant to people in the fields you will be working in. Try to avoid using abstractions that you feel you "have to" perform for disciplinary reasons. This imaginary is for you; write what feels authentic to you. Don't try to make your imaginary read like a formal proposal—it will needlessly constrain you. Enjoy writing what is on your mind and be open to options. When you relax into it, the process sparks curiosity and intuitive speculation.

As you do the following exercises in the space you've made for yourself, find time to experiment with perceptive stillness in which you just sit and think with what you've written. Let yourself slow down and allow reactive impulses to cool off so that there's more openness to the process. This helps you come into a place of nonreactivity so that you can identify, refine, and transfigure your own reflective capacities. Ultimately, spaciousness and stillness help release fixed ideas about how our projects should be, relaxing us into uncertainty and surprise. This helps us connect ways to work with our curiosity as well as our dreams about worlds we want to live in.

Example 1.1 provides Tariq Rahman's research imaginary. It represents four days of responding to the prompts below. Notice how Rahman takes a free and breezy tone to his writing. It is purposely informal and leaves

EXAMPLE 1.1 Rahman's Research Imaginary

When you go to big Pakistani cities like Lahore, you can see that about half of the population lives in *kachi abadis*. These are informal settlements in the interstices and outer edges of the country's cities. The other half live in mohallas (neighborhoods) on legally legible "plots" of land which used to be bought and sold between families as part of everyday life changes, such as converting land inheritance into cash or partitioning spaces for new family dwellings. But now these are highly valued global commodities. Urban plots in Pakistan have become a heavily financialized property market, as a local and diasporic network of investors vigorously trade plots as capital and for legal and illegal development projects.

So, I'm imagining that I need to see how Pakistan's economic system is emerging in a time marked by the ongoing war on terror and global finance. These globally financialized plots open up a way to look at what land is becoming in Muslim countries post-9/11. Plots seem like changing forms of land that reveal how housing and property is administered, how global banking and markets occupy and transform Pakistani family lands, and how financial rationalities are emerging in Pakistan.

Broadly conceived, I imagine my project is about the way that finance shapes, and is shaped by, Pakistani cities. On the one hand, I am interested in the role that financial practices play in determining urban landscapes and life in Pakistan and the new forms that rizq (material sustenance given by the divine) is taking in a global finance context. On the other hand, I am interested in the particular form that these practices take in the context of Pakistan's property market, which is socially networked, technologically mediated, material, and transnational in some rather unique ways. Even despite the entrance of Pakistani land into the global markets, you can see people making these trades via personal networks, kinship relations, and various social media platforms like Google Earth. Sometimes this is about building homes, sometimes this is about how people enter risky worlds of speculation.

If I imagine fieldwork, I think of building on what I did last summer. During that time, I primarily worked through two channels: an acquaintance who lived in a local mohalla and the office of a property portal. Mohalla literally means neighborhood, but specifically refers to a traditional form of dense residential housing, rows of which stretch out in organically developed, winding directions, often separated by narrow, unpaved lanes. A property portal is simply a web space where property available for sale is listed. I want to speak with people impacted by this new plot trading schema: mohalla residents, kachi abadi residents, real estate agents in Pakistan, and elsewhere, financial investors, people who work as a *stamp farosh* (a kind of notary for property transactions), and *patwaris* (the government-appointed record keeper for properties in a given area) who work with records and documents.

To be sure, Lahore's rich history of urban planning and contemporary urban dynamics would make it a fascinating place to anchor my project. As a Pakistani born and raised in the United States, I am never sure about where I stand with respect to risks to foreigners in Pakistan. I've gotten to the point where I feel relatively safe in cities such as Lahore, but Balochistan (the province Gwadar belongs to) in particular has been a hotspot for bombings and kidnappings in the past. Having conducted fieldwork in less "stable" parts of Pakistan before, I'm aware of the toll this can take on one's personal well-being as well as research, and I'm not sure that I want to put myself in that position again.

I should add that another option is a multisited project, which could base itself in Lahore, Gwadar, and a smaller city in rural Pakistan, where the effects of both the property market and the China-Pakistan Economic Corridor are increasingly present. This might make the project a bit thin with respect to place, but could also broaden it in terms of theory, which sits fine with me. As the themes I'm interested in are incredibly translocal (the Lahore developer who is building a housing scheme in Gwadar, for example), such an approach might even be a necessary one.

Investors, who make plots global, are quite scattered as they are typically working professionals from around Pakistan as well as the rest of the world. One of the places they do gather, though, is WhatsApp groups. What participant observation can mean in these spaces is something I'm still trying to figure out. Also, there are literally thousands of these groups, and I'm trying to get a better sense of the scale of this world in order to best situate myself in it ethnographically. Spending time in the offices of developers and agents is probably what I'm most excited, as well as most anxious about. My aim is to essentially find one or two of each of these figures that will maybe allow me to be around while they conduct their business. But they might have trade secrets they don't want people to know about. As odd as it feels to write this, the idea of experiencing the mundane, day-to-day labor of making property investable absolutely thrills me. The developer and the agent that I had the most fruitful conversation with and was able to visit on more than one occasion were also people that I could naturally laugh, philosophize, and complain with.

The property portal is another site where I imagine that participant observation can be valuable. However, my access there is through a family relation, and I'm still grappling with how to handle that. On the one hand, when it comes to producing investability, there is no single actor in the market that is more significant; the company's property index, blog, and news stories are the closest thing that exist to expertise. On the other hand, my view of the site will almost certainly be critical. And while this family relation shares many of my theoretical interests, including my concerns about the market, it is clear that they would not want the company to be represented in a negative light or associated with anything critical of Pakistan more broadly, wishes that I do not want to violate. Anonymity is of course possible, but given the company's unique position in the market, it will significantly restrict how I can write about my time there.

Things feel very liquid right now, but broadly speaking, what I think will be unavailable to anthropology as a discipline if I don't do this project will be bringing together the novel ways that land becomes financial with the ways finance manifests and flows in postcolonial and post-9/11 contexts. The concepts I'm trying to bring together are global finance and urban land and family property and media technologies, sort of confluences of material and immaterial forces that are converging and reshaping each other from within and outside of Pakistan.

lots of open-ended space for multidimensioning. Yet, at the same time, he covers all the project possibilities that he could think of per the prompts we provide.

This example is meant to inspire you. You will have your own approach and style of writing. Note his use of terms pertinent to the social worlds he is working in, such as mohallas, kachi abadis, and rizq.

Day 1. Write Your First Research Imaginary

Drawing on the writing process form you decided on in Interlude 1, on a blank page (physical or digital) write the title "Research Imaginary." Then read these simple prompts, one by one, and write what arises. And don't overthink! These prompts are not questions to answer directly—they are meant to stimulate your thought process before you write. In other words, the research imaginary is a flowing narrative of what comes to your mind after you read the prompts. Try to keep the first iteration to 1,000 words (two pages), single spaced.

1. *Clearly state what you want to study and why.* What is the large emerging and compelling process that you want to understand and that you want to engage in the world? To prepare to address this effectively, contemplate what would be lost to ethnographic literature if your project isn't done. This will help you write a few sentences about why this project is important to scholarly and social worlds. Again, don't worry about whether that "why" is a formally justifiable reason—just summarize what you and others might know about this topic at the moment and what you and others may not know about. State why you, as a researcher and person, think the project should be done. Focus on possibilities and potentials—what your project might offer to the world.

2. *Write in detail about what you want to do in the field, and use ethnographically relevant terms.* Here are prompts for details you can weave throughout your paragraphs. Again: *do not* write a paragraph for each prompt! Instead, hold these items in mind as you write about what you *need to engage with, experience, inquire of others, and understand.* As you saw in Example 1.1, consider using terms relevant to the languages and experiences of those you will be working with.

   a. Clearly describe the *specific processes and objects that you want to investigate.* Include some information about the broader social, political, and spiritual contexts in which they are unfolding.

   b. Provide details about *who* (human and more-than-human) and *what* (things, technologies, documents, and other material stuff) are involved in these processes. Note how you imagine accompanying people and other beings and how you will engage with their experiences and the *material parts of their lives* firsthand.

   c. Write about the *specific spaces and sites* that these processes are occurring in and how you will gain access to them (transportation, access, permissions).

   d. Be sure to include the *social and material resources* you will need to make your research and everyday life possible.

Now, put the imaginary away for a day. We're serious. Do not look at it again for about twenty-four hours. Think of it as bread or yogurt—it needs to rest and mellow in order to take shape, gain flavor, and begin a life of its own. This one-day pause holds the spaciousness and stillness and helps you return, when ready, to a state of perceptive openness.

Day 2. Identify Project Concepts within Your Imaginary

Now that you've freely written your imaginary, it's time to surface its key concepts. These are the vital conceptual building blocks of your project. We focus on two kinds of key concepts in this module: the 4Ps and broader contexts. The 4Ps are porous groupings of *people, parts, processes,* and *places* that you need to directly engage as an ethnographer. The 4Ps help you identify and connect with the project's broader contexts. Broad-context concepts are so broad that people either don't often directly experience them, or when they do, it is a concept that generalizes a state or experience shared in general. These broad contextual concepts actively shape the spatial and processural flows and frameworks of lived experiences.

You may have already noticed that we prompted you to include these project aspects in your imaginary free-write. All in all, they compose the main empirical aspects of an ethnographic project; you can think of them as the essential anatomy for anthropological field research—what brings the project to life. Again, these concepts may be rendered in languages relevant to those with whom you will be working. Table 1.1 explains each of these key concept groupings.

TABLE 1.1 Key Concepts Explained

4Ps	The main empirical aspects of a project; what you will directly experience in the field.
People	The *people* are entities who make up the social field that researchers engage; this term opens to all varieties of persons and beings that ethnographers and their interlocutors engage, whether human or more-than-human. This could include, for example, mothers, children, grandparents, farmers, bankers, dogs, ancestor rivers, spirit helpers.

Parts (things)	The *parts* are all the materials or objects that dynamically interface or interconnect with people. This could include iron, air, water, voting machines, shovels, bombs, cells, boundary markers, lumber, documents, and sinkholes.
Processes	*Processes* are the actions, procedures, or changes that are made possible by the interactions of people and parts. Importantly: these are processes you can and must directly experience within your field settings. This could include farming, hunting, eating, migrating, manufacturing, real estate transactions, remote sensing, conceptualizing, singing, raising children, making wills, intimacy, incarceration, voting, dying, healing, and memorializing. Key processes exhibit these important project-defining predicaments, problems, paradoxes, pressures, and abuse or use of power.
Places	*Places* are the physical and conceptual terrain and sites where people, parts, and processes are made possible. This could include homes, a particular village, villages in general, a country, a nation, an office, a shop, the marketplace, a site, the lunar surface, the domains of spirits and ancestors.
Broad Contexts	The *broad context* concepts are terms that you *and* interlocutors use to indicate the most extensive spaces, forms, or processes within which—or against which—people experience their lives and relations. These spaces and processes are typically too big to directly experience, but they are being enacted at a scope that creates the backgrounds for social life. This could be nationalization, global pandemics, dispossession, economic marginalization, free trade regimes, religious conflict, wars against terror, climate change, neoliberalism, capitalism, planetarity, a particular war, industrialization, financialization, political regime shifts.

Our prompts for surfacing these concept groupings in your imaginary are not about sorting things into hard-bounded positivistic categories. Instead, our prompts help you pay attention to how project concepts relate and coexist within and between porous groupings, such as a person who is involved in one or more processes, or a process involving one or more persons. We use these porous conceptual groupings throughout the handbook, so please mark this table to return to if you need a definitional refresher.

For the next steps, you need a color highlighter system, physical or digital, depending on whether you'll be working with your research imaginary on a piece of paper or on-screen:

1. *Read the category descriptions carefully.* Pick five highlighter colors: one for each of the 4Ps and one for the broad contexts.

2. *Color-code the 4Ps and the broad contexts.* Read your imaginary carefully and think about every word. Get ready to deploy your highlighters to mark the 4P and broad-context concepts: *as single terms, phrases, or, if necessary (such as for processes) full sentences.* Do the highlighting one concept grouping at a time, according to the following guide and examples. Be careful and selective about whether the concepts you identify are vitally definitive concepts (terms or phrases) for your project, as we will impose a limit on the number you settle on.

   – **4Ps**: the *people, processes, parts,* and *places* you want to directly interact with as part of your data gathering and your personal relational and visionary commitments.

EXAMPLE 1.2 Identifying the 4Ps

> Your project is on "voting and race" so you have to detail the *people* who are involved (such as voters, poll workers, election commissioners, or grassroots organizers). In the category of *people,* your project may also include entities other-than-human who in other social worlds are understood to matter or deserve political representation, such as mountains, rivers, and ancestors. Other elements would be the *parts* or things that matter (such as voting machines or polling lists or identity cards), the *places* where things are happening (such as polling places or campaign offices or buildings in which communities organize), and the *processes* like voter registration or community organizing that matter to the people you're working with and will be perceiving in the field. You will also need to detail the *parts* of those processes, like registration applications, registration venues, online and in-person registration tools, and reasons for registration disqualification. You are

interested in such processes and parts because you envision transformations of political processes, perhaps in ways that abolish voting in its current form or completely. In this statement "abolition" arises as a *process* concept that may reshape the parameters of what you envisioned your project to be or do, leading to multidimensional design possibilities.

– **Broad Contexts**: the overarching and through-putting social, political, environmental, and structural conditions that create the context for your project and shape people's experiences of the processes that you are investigating.

EXAMPLE 1.3 Identifying Broad Contexts

Being a voter or the process of voting may occur within the *broader contexts* of labor-practice inequalities (that prevent people from leaving work to vote), gerrymandering, gentrification, gender inequality, racial disparity, legal and illegal forms of voter suppression, and the large-scale adoption of new voter technologies.

3. *Review*. Read through once again and highlight as necessary.

When you are finished, put the imaginary away and take another day-long break.

Day 3. Reflect on Project Congruence, Rewrite Your Imaginary, and Engage Multidimensionality

You are going to write a concept-work reflection and then revise your imaginary. This is the first of many reflections you will write to help you iterate your multidimensional concept work. Reflections are an essential aspect of project listening. Throughout this handbook, we give you prompts to help you to go back and assess your work. To do such reflections well, we advise that you take time away from the original work to clear and open your mind.

Some of the best reflection happens after you unhitch from the engine of productivity and find a way to lightly detach from what you've done so far. The reflections we outline in this handbook have multiple purposes: to explain how and why you did what you did, to question what you did, and to create a space for imaginative "next steps," speculations, and intentions.

How do you start writing a reflection? We always provide formal prompts in the modules, as we do here. You can handwrite your reflection, but you must eventually write it in a digital format that you can share with others and archive. Before you begin, try to be rested, hydrated, and well fed. Caring for your body helps support good concept work.

As you move through the reflections, beginning with this one, you will notice that we ask you to assess your underexamined *assumptions* (experience-based or externally given knowledge) about project concepts, which are different from noting and exploring your *intuitions*. Assumptions are automatic ideas, either informed by your own singular experiential events or by what you have been given to think by others. Assumptions can leap over intuitions, in the sense that they can serve as preformed blockages that prevent your attunement to the nuances of other factors and new information you are perceiving in your mind-body as you imagine, read, process with others, and engage with your preliminary and later fieldwork. Being clear about when you're making assumptions versus exploring intuitive speculations can help you be open to possible insightful and exciting breakthroughs, and elaborate where needed.

This first reflection and subsequent revision focus on project *congruence*. Remember, by congruence we mean that the project aims and concepts hang together well. We'll work on congruency throughout the handbook. In these initial project congruency prompts, you will explore your understanding of the relationship between your own life and goals as a researcher, your assumptions about fieldwork relations, your selection and balance of key concepts, your understandings of your project's conceptual relationships, and how your project situates within broader worlds. We suggest you read through the prompts before beginning.

Refer to your highlighted research imaginary as you reflect on each of the following prompts:

1. *What is becoming clear about your personal vision?* Now that you've written an imaginary, take a moment to reflect on how

the project expresses your experience or motivation. Consider these prompts:

a. How did you come to this project and why does it motivate you? What is your social and political positionality with respect to this project?

b. Does this project reflect personal and shared collective aspects of your life? Why might this facilitate the project or be an obstacle?

2. *Is there a balance of people, places, processes, parts, and broad contexts?* When you look at your highlighted imaginary, do you see more of one color than another? Write about any imbalances you see—more people than processes, more processes than people, more places than parts, more broad contexts than processes. This last one is common—designers describe large contextual processes but omit the exact processes they need to observe, find out about, or participate in during fieldwork.

3. *Are the connections among people, places, processes, parts, and broad contexts clear?* That is, have you made it clear how the people, processes, parts, places, and other empirical features of your imaginary relate to broader contexts or other theoretical concepts? What specific connections you are making about them? Are you assuming that these connections are accurate? What do you need to do to investigate your assumptions?

4. *What are you assuming about your own relationship with the people, places, processes, parts, and broad contexts you describe?* Now is the time to slow down and consider your project's conceptual and fieldwork assumptions. For every instance in which you write that you will go somewhere, have access to a space, do or observe an activity, interact with people, or gather information, take a moment to examine your social, political, logistical, conceptual, and ethical assumptions. Try to extend this assessment as far into your ethnographic imaginary as you can, asking as deeply and openly about your assumptions as possible.

EXAMPLE 1.4 Addressing Assumptions about People

If you wrote, "I will go into clinics and interview adolescent patients," ask yourself, "Is this something I intuit might be possible, based on confirmed information about how to access clinics?" Or is this an unexamined assumption about the space, people you want to connect with, and their availability? Address the nature of the assumption, such as "I assume that I can go into clinics and interview adolescent patients, but I need to know whether this is allowed by my institution or theirs, or whether they would even want to talk to me in a clinical versus other setting." Take a moment to go further with the problem of assumptions. For example, address these questions: Will you interview only adolescent patients to understand how the adolescent patient is made? Who else might be important to interview, in and outside hospital institutions, when it comes to the concept of "the adolescent patient"? When you imagine your project in this way, you are beginning multidimensional concept work.

5. *What are you assuming about the relationship between your reason for doing the project and what you plan to do? What intuitions do you have about that relationship that you can explore further?* Following earlier prompts, spend time exploring your assumptions about what you think "needs to be done and why" and what you're assuming about the structure, feasibility, and ethical considerations of your fieldwork. At this point you're not looking for solutions, just acknowledgments about what you might be assuming and how certain dynamics and logistics need to be considered when designing your project.

EXAMPLE 1.5 Addressing Assumptions about People, Parts, and Processes

If you wrote, "I will interview physicians to find out why they chose to use a particular therapy (a project "part") for adolescents," ask yourself if this is really how you will know "why" they make these choices. Consider if other sites, subjects, or methods would better answer your question. You can even pause to acknowledge if you are confused about *how* finding and accessing sites, subjects, and processes works. Add a few phrases or sentences to address these problems. For example, "I assume that by interviewing physicians I'll understand why they chose a particular therapy, but I'm intuiting that this would make me dependent on their discourse rather than watching them choose while they work. But observation might put patients at risk of losing their confidentiality, so I need to explore this further."

6. *What are your intuitions about what you need to know to deepen your knowledge about the project?* Do you need, for example, more familiarity with the field, with previous research, with other kinds of research, with your own relationship to the project? Do you need to know the history of peoples, places, parts, processes, and contexts better? Add a few short sentences or phrases here and there to indicate these needs.

7. *Which people, places, processes, parts, and broad contexts seem disconnected from each other or connected to processes or contexts* beyond *your imaginary?* List concepts (terms, phrases) that seem disconnected or less connected within your project. These are not problematic; in fact, they are often openings to multidimensionality that you will begin to explore in the succeeding modules.

8. *Rewrite your imaginary.* Now, using your responses to these questions, revise your imaginary and judiciously add information and sentences where needed. Note any thoughts about project congruence and disconnection. Keep this iteration to 1,200 words (about 2.5 pages).

Next, follow the prompts to add a final paragraph to your imaginary that assesses your project's current and potential multidimensionality. These prompts are designed to help you begin to perceive which kinds of people, places, processes, parts, places, and broad contexts you are bringing together in new ways. Think about what potentials your project has to combine concepts in ways that generate new understandings of emerging or changing social processes. This is just a beginning step toward designing multidimensionally, so don't feel that you have to have all the answers now!

Follow these prompts to write a short paragraph.

9. *Imagine project multidimensionality.* See how you can open to the spirit of multidimensionality: creating a project that puts concepts together in new ways. You may or may not feel that your project is doing this yet, or even that you understand what it could be. So far, you may be creating an imaginary that is based on a project that someone else has done, but in a different place with different people; in other words, your project may look like a case study so far. That's all right! Write notes for each of these prompts. If you cannot answer prompt a, then just answer prompt b.

   a. *So far, what do you imagine is unique and interesting and unusual about how your project is bringing concepts together (people, processes, parts, places, and broad contexts)?* Your sense of this may come from your own experience, the needs of your communities, or from familiarity with other people's work on your topic. If you're having trouble finding this uniqueness, go back to item 5 above to draw inspiration from concepts that seem disconnected or connected to processes outside your project. Speculate whether these concepts might open your project to some unknown opportunity to bring people, parts, processes, places, or broad concepts together in new ways.

   b. *What do you think you need to do to find out how to make your project more conceptually multidimensional?* What might help you identify new concepts or be clearer about those you are bringing together in new ways?

10. *Write a new, final paragraph for your imaginary.* Take these notes and write a short paragraph. Adding this paragraph will add another 250 to 300 words to your imaginary; keep your total at or fewer than 1,500 words.

As you did with the first iteration, wait at least twenty-four hours before doing the next exercise that finalizes your imaginary.

Day 4. Read, Reflect, and Make a Final Revision

Doing the following process of slow reading and overall reflection is necessary before moving to the next phase of finalizing your key concepts. This slow reading is also an act of careful listening—how does your research imaginary strike you now that you've done the nuts and bolts of assessing your conceptual framework? Here are some prompts:

1. *Read your work.*

    a. *Read with curiosity and interest.* This is the first step in cultivating an intuition for the research project and multidimensional design.

    b. *Perceive your embodied responses.* Track how your whole body is reacting to what you have written. Know that your bodily reactions are simply providing information. We take this information seriously because we consistently find that researchers always go through emotional highs and lows while doing this work. Listening to reactivity without making the researcher or the project "wrong" helps to cultivate emotional balance, important for all concept work. When you can listen to your project in this nonreactive and even-minded way, you'll widen the conceptual mind space for more insight and connections to arise.

2. *Reflect on your relationship to the project.* Write answers to the following prompts. Go slowly and answer as thoroughly as you can. See if you can reflect without shaming or dismissing what is there and what might need to be changed.

    a. *How do you feel about your imaginary overall?* Where in the narrative do you feel elated, anxious, confused, or irritated? Or where do you have no feelings at all? When there is elation, cultivate that excitement; maybe even free-write to see what other insights arise. When there is dread or confusion, see if you can allow those emotions to simply be indicators that something isn't quite fully formed, which takes time. You can write freely about that, too.

b. *How do you feel about the emerging sense of project purpose?* What is the relationship between why you think scholarship needs this project and why you want to do it? Which of these justifications do you feel least clear about and why? What do you need to do to make each justification clearer?

c. *How do you feel about your situated, personal vision?* Does it reflect the range of your experiential and intuitive senses of what the project means to you? Do you feel like you can share a sense of the possibilities you are committed to as an ethnographer and as a member of communities? Do you wish to share parts of this imaginary with a group or keep it more private?

d. *How do you feel about the multidimensional potentialities of your project?* Just reflect on how you feel about finding ways to bring new concepts together. Go easy on yourself—Modules 2 to 4 will help you get there!

e. *What needs more attention?* Where do you see the strengths and clarities, and where do you see gaps and problems?

After writing your reflection, the next phase is to revise the imaginary as needed. You can do this on your own or get a partner and talk through each other's work. Here are some tools to help you:

3. *Do a final revision of your imaginary.*

a. *Make changes.* Return to your responses to the prompts in the Day 4 reflection to help you take things out, add details, and clarify.

b. *Be judicious.* Be careful and do not overburden your document: do not add more than half a page, or five hundred words.

All in all, at the end of these four days, you should have an approximately 1,500-word research imaginary. If it is longer, see whether you can edit it down to 1,500 words. Any longer than this is unwieldy; any shorter is probably incomplete.

EXERCISE 1B Create Your Key Concepts Table

Based on your highlighting work and reflective edits to your imaginary, it's time to begin creating a Key Concepts table. Why a table? A table allows you to productively contract your project into a different shape after the expansive linear narrative act of imagining. What do we mean by *key*? At this point we mean no more than thirty-five salient terms or one- or two-word phrases that define the project and its processes.

Before you begin, look at an example of what the table looks like and how you can fill it out. You may remember, from the introduction, that Tariq Rahman proposed to research "liquid land" (his multidimensional object, or MO) in Lahore. He got to this MO by following his interest in how new real estate plot markets in Pakistan are financialized within global networks, with "land" gesturing to larger questions about sovereignty, movements of bodies and money across boundaries. Example 1.6 provides an example of his early work to identify key concepts.

EXAMPLE 1.6 Rahman's Key Concepts

4Ps	*People*: real estate agents, plot owners, kachi abadi residents, developers, stamp faroshi, patwaris, investors *Parts (things)*: real estate deeds, paper files, digital files, transfer forms, WhatsApp communication platform *Processes*: real estate trading, Islamic investment, rizq, digitizing plots, urban planning *Places*: plots, Lahore, kachi abadis, mohallas, cities, informal settlements, city bureaucracies
Broad Contexts	Global real estate trading, the Pakistani state, war on terror

The template for the Key Concepts table is provided in Table 1.2. Follow these instructions for filling in the table.

TABLE 1.2 Key Concepts

4Ps	*People:*
	Parts (things):
	Processes:
	Places:
Broad Contexts	

1. *Go back to your highlighted text and fill in the table with up to thirty-five terms or phrases, keeping an eye on balance.* The goal here is to clearly see the project concepts that will drive the eventual development of a cogent research topic description and then a cogent MO that represents how the project hangs together. You will come back to this table often.

2. *If you are having trouble making decisions about what to place in the table, ask yourself the following questions:* What concepts are necessary to convey the who, what, where, when, and even how of this project? Don't worry if what results seems like a laundry list—a jumble of things that seem connected but also disconnected. This is a conceptual sketch of your project as it begins to reveal its unique shape and features. The "what do these have to do with one another?" connection work will come later.

This kind of work may seem reductionist, but it is not: you are using your impassioned narrative to create a feel for the whole project in a way that will give it a broad but manageable scope. We have found that it cultivates an intuitive feel for a project, where intuition helps ensure that you keep key concepts in the frame and maintain balance and coherence among them. It also allows you to surface concepts that lead to breakthrough hunches that create exciting research designs. Cultivating intuition in this way hones the multidimensional potential of project design.

You're welcome to work and rework this table for a few days—but don't make it unwieldy. Keep all your iterations so that you can refer to them as necessary.

You will work on your Key Concepts table again in Module 2. It will help you to situate your project to the writing, debates, and questions (or silences) of your project concepts—namely, the topics and concepts that structure your project's theoretical and empirical aims. You'll explore where your project easily fits—and doesn't fit!—into literature categories in order to see some further ways to add new conceptual and topical dimensionality.

Now on to another key table: your project grid.

EXERCISE 1C **Research Project Grid 1**

Create a document with a landscape layout and reproduce the grid in Table 1.3. Fill in the 4Ps and Broad Contexts concepts in the rows found in the *top* section of the grid. Leave the Specific 4Ps column blank for now.

TABLE 1.3 Research Project Grid

Research Description
Multidimensional Object
Scoping Question
Significance Statement (reduce to 50 words)
Key Concepts **4Ps:** **Broad Contexts:** **Other Concepts:**

Key Literatures				
Process Clusters	**Data Sets**	**Data-Gathering Questions**	**Specific 4Ps** *who/what/ where/which processes will you engage to answer this question?*	**Methods** *how will you get the data?*
Add rows as needed . . .				

You will fill in the remaining parts of the table as you complete the modules, but along the way, note how they will relate to each other and form project congruence. The Research Project Grid table serves as a flexible frame for your multidimensional concept work and as a shareable, one-page snapshot of your project. It is an excellent tool for developing and assessing project balance, by which we mean the way that project concepts connect and build off each other, and for the ways in which you communicate and explain such connections. For example, the project grid helps you explain how the subjects you will interact with, and the methods you will use, directly connect to your data-gathering questions. It will also allow you to see how, in turn, your data-gathering questions are connected to your literatures and, ultimately, to the main theoretical research question (scoping question) you have for the whole project. In addition, you will identify the overall significance of the project, which helps to make its worth justifiable to others.

As a portable quick-view, the Research Project Grid table also provides a clear focal point for work with collaborators, advisers, and funders. You may refine the grid categories to suit the needs of your project, but note that this tool should allow you to maintain your focus on the project's conceptual and dimensional coherence.

As you familiarize yourself with the grid, imagine that you are creating a three-dimensional tensegrity mobile of concepts: an object made of interconnected concepts of various shapes, sizes, textures, and colors that must hang together in a complementary arrangement. As you complete the modules, we will coach you on how to continue to fill in the grid.

Collective Concept Workspace 1

The exercises for this module are about slowing down, pausing, and learning to intuit the micro steps of concept work, in addition to getting initial clarity on your project design. All of it matters—from developing the 4Ps and broad-context concepts to stepping back to create spaciousness in your thinking. A collective concept workspace is convened by communities of researchers after agreements on process are made (see interlude 1) and the exercises in each module are completed by all. This amounts to a total of ten to eleven meetings, depending on how researchers want to engage with each other after Module 9.

The discussion prompts here will help you and your group reflect on this (perhaps unfamiliar) process, and they will surface the specifics of any project confusion and clarity. Moreover, they help to build intuition for your project design. You can consider these prompts on your own, but they are meant to be answered and discussed in a group setting. Your group can also decide on its own prompts. After you have completed group work, take a break for twenty-four hours. Then reread your imaginary and consider revising it based on the feedback you received. Remember that this is very initial concept work that is in flux and most likely lacks certainty. So try to let go of *getting things right* and just allow the process to surface. With further concept work, you will continue to refine the conceptualization and design of your project.

Reviewing your agreements before you begin every collective concept workspace can serve as an opening ritual, which can strongly support the foundations of conscientious, collective work. It also can reveal what's working and not working in group dynamics. The agreements, therefore, can be adjusted to facilitate group-work processing as well as offering care and support to everyone in the group.

1. How did it feel to build spaciousness in the concept-work process?

2. How did it feel to identify the 4Ps and broad-context concepts? How does it feel to see them in the Research Project Grid table?

Provide as much detail as possible about them, and imagine how they will be operationalized in ethnographic work.

3. How did it feel to reflect and listen to the research imaginary? Any new insights?

4. What assumptions are you making about the project? And how do you reflect on those assumptions? Does it feel like it is congruent or not so much?

5. What connections are you making between the project concepts? Where is there balance and imbalance in terms of people, places, parts, processes, and big-picture concepts?

6. When you look at your Key Concepts table, what feels right *in your body* when you read through the table? What areas or connections are fuzzy or confused? Fuzzy doesn't mean "bad": it can be the place where real multidimensional vigor emerges!

7. What feels clear and exciting about the project?

8. What breakthroughs would you like to have?

MODULE 2

Focus on Literatures

Now that you've imagined your project's processes and identified its key concepts, it's time to situate your emerging project idea within a larger conversational milieu. This milieu contains literatures: bodies of categorically distinguished work, conventionally bounded by discipline, history, topic, and theme. This definition comes from the US context, where we trained and have also trained students. We recognize that what counts as a conversation, literature, or shared disciplines—and what constitutes their boundaries—varies greatly across discipline, place, and region.

We encourage you to engage this literature concept-work process with respect to your own practices for interacting with others' works and knowledges. The overall goal of this module is to help you effectively navigate a lot of already-done work that is relevant to your project. It's important to understand that we call that work "literatures" not because all work relevant to your project is textual, but because ethnography's multimodal forms are still generally anchored by texts that engage with other texts. That being said, our process can also be extended to working with more-than-textual cultural and social forms. Engaging with others' work carefully is an important aspect of research's meaningfulness and politics.

Link project concepts found in the research imaginary to three project-defining literatures.

Most researchers want to contribute to literatures. They often first make this kind of claim in a research proposal: "this project brings together the literature on *x* with the literature on *y* in a novel way." Making these claims about how a project is situated in literatures often comes after lots of trial and error, which is unavoidable. This module sets the stage for creatively identifying and juxtaposing literatures in a more deliberately mindful way. In other words, it introduces you to working with literatures in ways that may lead to unusual but fruitful literature combinations—a process we take you further into in Module 4. We have found that creative literature work is also important for designing research with respect to otherwise practices of knowing and inquiry. Consciously juxtaposing bodies of literature in unusual ways can soften hard-bounded ideas about what counts as a valuable project and what kind of data gathering and social interactions it requires.

The normative literature categories typically deployed in social science research designs tend to be products of knowledge siloing that emanate out of Cartesian epistemologies, Enlightenment reasoning, and European imperialism. As Ngũgĩ wa Thiong'o and others show, those knowledge-making products are often still entrenched in the authoritative domains of professional organizations and academic departments.[1] In US-based academies, literature categories are often connected to disciplinary productions like publishing categories, journal structures, research foci, job titles, and courses. However, today's literature categories can also reflect changes to these normative institutional structures. See, for example, the anthropology of labor, visual anthropology, anthropology of religion, the anthropology of technology, Latin American anthropology, and also crosscutting categories such as queer anthropology and Indigenous studies.

All of these literature categories have theoretical, methodological, and ideological dimensions that make them meaningful in societies at large. If you work in an academic setting, we recognize how important it is to make your research legible within recognized categories of scholarship. This may be for the instrumental reason of getting funded. But they can also be done to connect with others doing transformative work—for example, with people participating in anticolonial and abolitionist efforts that make knowledge forms more experientially accurate, lively, and effective and that foster new conversations among broader communities of knowers and inquirers. The exercises in this module can help you to legibly situate your work within existing literature categories while at the same time stretching and breaking their boundaries.

The first exercises help you identify and justify three key literatures that make your project both legible and uniquely innovative. No matter how customary or uncommon your project's literature juxtapositions and conceptual constellations are, you will need to clearly define them and explain their relevance to your project's aims. This is a pivotal research design process. Every researcher we know, whether student or seasoned pro, has experienced project design setbacks simply because they did not locate their ideas within literatures early or clearly enough. Our approach to this saves you time and prevents (overmuch!) frustration. And like all the processes in this handbook, it is meant to be iterated.

The three key literature categories you will identify—two topically defined literatures and one area-defined literature—are derived from your key-concept discovery process. These will probably be the literatures that present themselves right away. Identifying them provides a broad intellectual space for the next modules, which include more focused exercises of concept mapping and drafting a succinct research project description. In Module 4, you will revisit the literature engagement process so that you can confirm and adjust your work in this module. This can result in either changing one of the literatures you have landed on here and/or adding another one. This adjustment and/or addition can give your project productive multidimensionality.

The last exercises in this module help you use your literatures to find additional key concepts. Some of these concepts will be unique to one of the literatures, and some will be shared between them. You may find in this literature work some additional key concepts that you can use to update your Key Concepts table. This table will be the basis for your concept map in Module 3. The last exercise in this module is to make a reading schedule that will help you get ready to write a narrative project proposal in the future.

As a process that helps organize literature engagement, this module is oriented to those just beginning to work with literatures as well as those who already have a strong grasp of them. For those starting out, doing all the exercises will help. To those who are familiar with how to work with literatures, resist the urge to skip over certain exercises or to skip the module entirely! The module is designed to help make clear the stakes and significance of the project, which is always useful, no matter how experienced you are. Doing all the exercises will add needed richness and clarity to your concept-work experience, eventually laying the groundwork for multidimensionality in later modules.

All of these exercises help you engage literatures as wholes, but we also want you to see *into* literatures as bodies of work defined by common and distinctive approaches to working with concepts.

Literatures and Their Concepts

Literatures are discrete conversations focused on concepts that include research topics as well as generalizable ideas. Topics are empirical or methodological foci oriented to places, peoples, processes, or things. Generalizable topics are commonly shared scholarly categories, such as anthropology, psychology, states, gender, sexuality, field methods, economy, law, water, or ontology. Literature categories are internally bound by shared attention to *common concepts*, be they specific topics (Alaska or nuclear weapons) or abstract ideas or processes (anthropology or ontology). However, not all concepts end up as bodies of literature. As we indicated earlier, the concepts that define literatures are often attached to other social formations like disciplines or journals. These literatures are often bridged by a shared concern with particular concepts, such as "being." So, while "being" or "ontology" are concepts that many literatures address, they are not currently commonly recognizable literature categories in and of themselves. This is in part because journals of being or departments of ontology are not *yet* common, although they may be in the future!

Here is the takeaway: identifying the literatures your project will contribute to is important for writing well-organized literature reviews. But when it comes to research design, the most important aspect of a literature is its key concepts, how those concepts do or do not intersect with other literatures, and what relevance they have to your project's design and possibilities.

EXAMPLE 2.1 Locating Concepts within Literature Categories: "Finance"

Let's take the example of "finance," a concept that several literature categories pay close attention to. You will find many anthropological studies that make finance a topical focus of project design and data collection, because it is a modern process in which lives are enrolled in capitalist modes of monetary exchange and production. The conventional literature category that the topic of finance usually fits within is "economic anthropology." Finance might be divided into more specific topics in this literature, with empirically relevant subtopics like derivatives, microfinance, or arbitrage. However, financial*ization* is also an idea-in-action that scholars like to theorize about. Thus, as a mode of capturing and managing large sums of money and of future-making abstraction in capitalism-impacted arenas, "finance" appears as a concept within a variety of literatures, from political anthropology to the anthropology of kinship and the anthropology of art. Therefore, a term's importance across a variety of literatures should *not* lead you to conclude that it is a literature category in its own right. But if one of your project's key empirical and/or theoretical concepts is finance, it will be legible to a variety of literatures, and the key is figuring out which one or ones you should *and* want to locate it within.

How do you use the concepts you have found thus far to identify your project's key overarching literatures? The first step is to pay attention to those key concepts that are most crucial to your project. We'll help you make this helpful conceptual contraction. There is a chance that you already have a strong sense of what these most crucial key concepts are, and therefore you might know the literatures that you expect your project to situate within. Or you may know one literature for one concept but not be sure of the others. Either way, it's important to distill the major concepts that animate your project and explore how they relate to broad literatures in order to make a project that is relevant to researchers and others engaged in social exploration and inquiry.

The literatures you will identify may be those that you might be "expected" to include, such as economic anthropology if you're studying financial practices, but also others that are currently "unexpected," such as adding performance studies to a project on financial practices. "Unexpected"

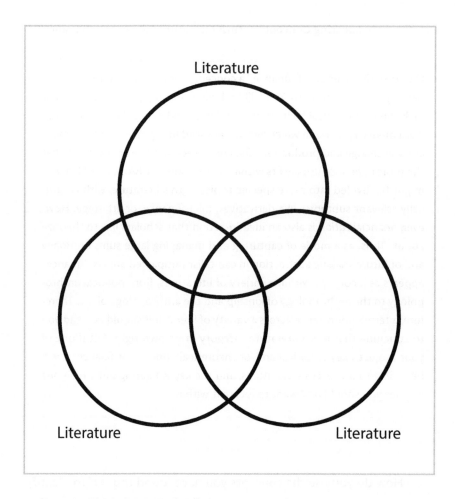

FIGURE 2.1 Literatures Venn Diagram

literatures usually emerge through the process of intuitively informed concept work, which might occur right away in this module or may happen later in the process.

The constellation of literatures you arrive at provides a framework for the heart of your project, most specifically for your multidimensional object (Module 6). Therefore, it is vital to iterate your literatures throughout the concept-work process. Managing your ongoing relationship with key literatures will certainly continue for the life of the project.

To begin, visualize literatures as discrete and overlapping topical and conceptual spaces. Our colleague Tom Boellstorff suggests that students view their literature-project relationship as a Venn diagram, as in Figure 2.1. The three circles in the diagram represent the three major literatures that your project will be shaped by and will contribute to, as well as their connections and disconnections. Your project design connects these literatures and will eventually join in those conversations. Creating a diagram like this, as you will in Exercise 2c, can help you to productively contract your project's scholarly focus, but open it, as we explain below, to multidimensional possibilities.

Included in your three literatures should be one area studies literature. We do not find it useful to think of "area" simply or only in terms of cartography. Rather, we understand it to be *generally spatial*: literatures defined by regional geography or geopolitical formations (South Asian studies, ocean studies), place (the anthropology of Vietnam, the anthropology of Mexico), or environmental spaces (watersheds, ecologies, ecosystems, outer spaces) are all examples of area studies literatures that are found across disciplines such as Indigenous studies, Black political thought, postcolonial studies, and anthropology. In addition to being generally spatial, we do not view "area" as fixed or something to be backgrounded. Just like a topically focused concept, area concepts are flexible, ever-changing, and meant to be theorized.

If your literature kinds and conventions are different from those we work with, we encourage you to adapt them to the following exercises.

When you complete your Venn diagram, it will help you recognize the integrative and expansive potential of your project, which is represented symbolically in the discrete and overlapping spaces of the diagram. This is another nonlinear activity, focused on perceived connections among your literatures. Note the *mandorla* (almond) shapes that appear as the circles overlap on the diagram. We consider these spaces to be fertile concept-work openings and you'll eventually populate them. For now, leave them blank so you can perceive them as open spaces of possibility. The discrete and overlapping spaces call attention to several dimensions: how project literatures stand apart from each other, how they share conceptual spaces with other literatures, and how they can open to new, as-yet-to-be-recognized areas of study and knowledge. We will build on this diagram in Modules 4 and 5 regarding finding multidimensionality because it conveys how literature engagement opens up possibilities for innovation.

How extensively will you be engaging literatures at this point? We do not expect you to do an exhaustive literature review, but we do have an exercise that requires a limited assessment of some existing literature reviews. We will coach you in how to derive the important topics and concepts your project needs to account for right away. This allows you to identify key literatures, which also will be useful in mapping your project (Module 3) and further describing its topic and multidimensional object (Modules 5 and 6). After these exercises, you'll make a reading schedule and *listen* to your literature work so that you can begin to articulate your project's scholarly significance. This is vital for scholarly inquiry and funding support.

Lastly, it is important to realize that some of these exercises may take a lot of time: expect to spend a week on them. Even so, you will find yourself coming back to them as you develop an intuitive and effortless feel for the stakes of literature engagement. And so, just like in Module 1, where spaciousness and perceptive stillness generated satisfying work production, make sure that you can identify when to pause, when to pull back, and when to keep going.

Exercises

EXERCISE 2A Use Your Project Concepts to Find Three Key Literatures

The goal here is to derive some key conceptual terms that will allow you to identify three key literatures that you want to engage and contribute to.

1. *Reread your research imaginary.* If you did group work after completing Module 1 (or you have since reflected on your work alone), consider whether you need to revise the imaginary more before you proceed. If you do this, be sure to take a break, ideally twenty-four hours, before you proceed to this next step.

2. *Review your Key Concepts table.* If you've redone your research imaginary, modify the table as necessary—maybe refining, changing, or even eliminating concepts. Remember that you do not need to have lots of words in these lists, just the key ones. Try to ensure that you have a maximum of thirty-five terms or phrases for the whole table. Taken as a whole, these terms will point in a variety of literary directions. However, the key literatures may not be visible yet.

3. *Find search terms among the key concepts.* Identify ten to fifteen salient terms in each of the two concepts categories that can serve as *search terms*. By this, we mean more or less common "keywords" that you might use to find studies and writings related to your project. When you search for those terms, use your gut instincts based on your project aims. But also bear in mind that you want to contribute to certain literatures. Know that you may add, change, or delete these terms as you go through this process. To illustrate, let's return to the example of Rahman's Key Concepts table (Example 2.2).

EXAMPLE 2.2 Rahman's Key Concepts

4Ps	*People*: real estate agents, plot owners, kachi abadi residents, developers, stamp faroshi, patwaris, investors *Parts (things)*: real estate deeds, paper files, digital files, transfer forms, WhatsApp communication platform *Processes*: real estate trading, Islamic investment, rizq, digitizing plots, urban planning *Places*: plots, Lahore, kachi abadis, mohallas, cities, informal settlements, city bureaucracies
Broad Contexts	Global real estate trading, the Pakistani state, war on terror

The aim is to consider which terms stand out as important "key" concepts that "cover" the more specific people, parts, places, processes, and contextual concepts. Look at Rahman's key concepts and consider the 4Ps box. *Real estate* is a key process that covers trading, buyers and sellers, buying and selling, residents, investment, and markets. Bureaucrats are also key state interlocutors. In addition, *cities* and specifically *Lahore* are the spaces that unite many of the project's fieldsites. In terms of context, the Pakistani state and the war on terror are overarching yet malleable forms, with *Pakistan* being the state in question. In terms of field-based processes, Rahman knew he would be positioning himself to witness specific legal, political, and even security-based activities within the broader context of global finance.

Pulling such terms out, we end up with a list of "key literature identification" terms to orient some online database searching to see what kinds of scholarly journals such terms appear in. At this point, we want you to simply search on individual terms to link them to the appropriate literature. Later, in Module 4, you will do another kind of searching: putting some of your key concept terms together to find out if (or how) other researchers have done similarly conceptualized research projects. This will help you figure out *how* your project will contribute to literatures. For now, you just want to see what kinds of literature categories your project *can* contribute to. Example 2.3 is a fully filled-out table using Rahman's project as an example.

EXAMPLE 2.3 Rahman's Literature Category Identification

Key Concept Categories	Search Terms	Literature Category
4Ps	Real estate	Economic anthropology
	Land tenure practices	Economic anthropology, legal anthropology
	Cities	Urban anthropology
	Kachi abadi residents	Urban anthropology, anthropology of Pakistan, anthropology of South Asia
	Lahore	Urban anthropology, anthropology of South Asia
	Bureaucracy	Political anthropology, economic anthropology, legal anthropology
	Investors	Economic anthropology
	Pakistan	Urban anthropology, anthropology of South Asia
Broad Contexts	Financialization	Economic anthropology
	Real estate trading	Economic anthropology
	Land tenure practices	Economic anthropology

4. *Begin to create your own Literature Category Identification table (see Table 2.1) by placing ten to fifteen terms in the Search Terms column.* Use five to seven maximum for each box that can act as keywords to use in search engines.

TABLE 2.1 Literature Category Identification

Key Concept Categories	Search Terms	Literature Category
4Ps		
Broad Contexts		

5. *Consider how these search terms can point you to your project's key literatures.* Using a scholarly search engine, search individual terms to notice the kinds of articles and books that each search returns. Take notes about the topics of the articles and books that appear. What are the general focuses of these pieces? What holds them together? For example, if the term is *real estate*, what kinds of journals does that topic appear in? Note the titles of specialty journals, such as *Political and Legal Anthropology*, or as articles appearing in edited volumes, such as a book about "trading land."

6. *Identify the major anthropological (or other) literatures.* Paying attention to the titles and journal types of the citations you retrieved, decide which one or two literature categories this term is most likely to be relevant to. Place those in the last column of Table 2.1. Your goal here is to explore what literature conversations your key concepts are likely located. This helps you

understand the major literature conversations you are aiming to join, and will help you later to articulate your projects multidimensional uniqueness and significance.

Let's return to Rahman's project to elaborate how to hone in on three key literature categories (see Example 2.3). Note that Rahman's key concepts map onto several kinds of literatures. For example, they map onto economic anthropology and also legal and political anthropology. But his research imaginary and key concepts are anchored in the empirical material of *urban land*; and the processes he is interested in are being enacted by people who are involved in how that land can and should be bought and sold as they negotiate in urban offices and homes in Pakistan and around the world. So for now the project must be understood to be connected to the politics and legal dimensions of *economic* processes. Economic anthropology and urban anthropology, therefore, are slightly more key to the project than legal and political anthropology, although he will also need to address those literatures when he writes literature reviews for his research proposals. Remember: for now, an important move is to contract the project into its three major literature conversations.

Let's take a closer look at the area literature possibilities. Like Rahman, you might see a variety of possibilities for how to name and frame your "area" literature. This will ultimately be determined by the way your project critically engages its main places. For example, cities are a spatial area category, so this strengthens his conviction that urban anthropology is a key literature. When it comes to place, the final literature grouping in Rahman's table could be the anthropology of Pakistan, or it could be postcolonial studies, anthropology of South Asia, or Pakistan studies. Each hails different disciplinary and interdisciplinary orientations. When it comes to "areas," consider how you express your project's locality and spatial parameters to be available for theorization and to be vital to the ways you work with and juxtapose the other literature. Given that there is a rise in articles focusing specifically on Pakistan and its cities and regions, the project will definitely contribute to Pakistan studies as it intersects with the anthropology of South Asia and postcolonial studies.

A final Key Literatures table for Rahman's project might look like Example 2.4.

EXAMPLE 2.4 Rahman's Key Literatures

Search Terms	Key Literature
Real estate, financialization, the state, land tenure practices	Economic anthropology
Cities, Lahore, Pakistan	Urban anthropology
Pakistan, South Asia, bureaucracies, land tenure practices	Pakistan studies

It's time to create this table for yourself, starting with the three literatures you have identified.

7. *Identify your key literatures (use Table 2.2 as a template).* Consider how your Literature Category Identification table surfaces a variety of categories. Consider those that your project is most positioned to contribute to and those that you are committed to contributing to. This may require a moment to pause and really listen to your project and your intentions. Select three that you feel—in your body—are your key literatures.

8. *Connect your search terms to your literatures.* Using Table 2.2, place each of your search terms next to the literature, or literatures, that they are relevant to, based on what you learned in your searches. You will find that concepts can be placed in one or more spots, sometimes in all three. This is how you begin to see how your concepts are—or should be!—relevant to major conversations in the literature.

   Relate your search terms to the literatures. Be sure to replicate concepts in the boxes if they are shared between literatures. You may find that you have to revise how you express your key

TABLE 2.2 Key Literatures

Search Terms	Key Literatures

concept terms in order to make them fit well. Be aware of any new conceptualizations that arise.

Now that you've identified your literatures and can visualize some distinctions and overlaps, you will turn to some preliminary literature reviewing in order to examine how your project engages those literatures and their conceptual concerns in ways that may not have been considered yet. In other words, it's time to imagine how your project can *specifically* contribute, conceptually, to the literatures you've identified. You may also begin to imagine how your project might include a literature, in the future, that is currently a mystery to you! You'll check that in Module 4.

Consider taking a twenty-four-hour break before continuing. As you do, notice your response to your work. Are you excited, agitated, overwhelmed, avoidant, or something else? See if you can listen to what's arising without judgment or reactivity. Practicing inhabiting equanimous spaces helps to cultivate a strong, intuitive feel for the project and its development. If you are feeling reactive or judgmental toward yourself or project, then see if it's possible to bring a feeling of friendliness and compassion to your experience. These tools go a long way toward supporting easeful mind-body states; they help discern the difference between intuition and assumptions, and they also can open insightful doors with project listening.

EXERCISE 2B Assess How Your Project Engages with Conversations within Key Literatures

In this exercise, you will use your version of Table 2.2 to assess where your project fits within the current conceptual landscape of the three key literatures you identified. It does so by helping identify how your project relates to concepts that represent persistent and emerging questions and directions within these literatures. This is necessary to design a project that is relevant to researchers and others engaged in forms of social exploration and inquiry.

The exercise also does more than identify additional key concepts for your project by prompting you to begin explaining *how* your project relates to these literatures. This will help you to structure a literature review for a grant, thesis, or dissertation proposal. Note that this exercise requires a lot of work and can be paced over a couple of days.

We start you on this road by suggesting reading resources that will allow you to engage with other scholars who have reviewed the literatures or key concepts that are relevant to your project. As you do this strategic review, don't panic if you find out that someone has done something similar to your project! This may not necessarily be a problem because the multidimensional concept work in upcoming modules can help distinguish your project from all others.

Sources for your review articles can include the *Annual Review of Anthropology* and other AR journals (like AR *Sociology*) or monographs and edited volumes that contain articles that introduce and review topics within a given subfield. Here's how to do this strategic review.

1. *Find one review article or chapter from an edited volume from the last five to ten years that is relevant to each of your key literatures* or *a vital concept crucially placed within those literatures. You will end up with a total of three articles.* A review article is a definitive analysis of years—even decades—of a whole scholarly discipline or subdiscpline (like economic anthropology or medical anthropology) of work in a particular discipline or subdisci-

pline, or on a particularly vital concept. Examples of this would be reviews of the field of economic anthropology or of the concept of political economy as an idea crucial to political anthropology. Here are some prompts to help you find these sources:

a. *Search in library catalogs or other databases.* Search these sources using your key literature categories or key concepts. If you don't find a current review of one or all of your key literature categories or concepts, refer to the search terms in Table 2.1 to identify one or two search terms that correspond to the missing literature. One such term might be *markets*, which, in Rahman's table, was a vital topic in two key literatures, economic anthropology or political economy. Use this process to identify keyword-based articles or chapters to review. Do the same search for your area literatures, finding reviews relevant to your primary discipline.

b. *Engage with literature conversations happening in your project's regional area.* It is vital to consider and include review articles by scholars who are located in the area (geographic, if that's what you include) where you are conducting research. For some scholars, this may seem obvious and easy to do, but for others it may require additional intentional research. Wherever your research is taking place, it's important to get to know the topics, questions, and debates happening in the local scholarly community. The conceptual terms pertinent to these debates may not be included in dominant knowledge production outlets, such as journals or databases. Therefore, over time, it's critical to get to know the publishing landscapes and patterns of your current or future colleagues, with whom you'll be in dialogue on your research topic.

2. *Read your three articles or chapters carefully.* Be sure to review the footnotes and especially the bibliography. Current literature or topical review bibliographies are sources of canonical, critical, counterhegemonic, and newly relevant sources.

3. *Write a reflection.* For *each* review article or chapter you find, write a response to each of the following questions. Be thorough and specific in your responses.

    a. *What is the article's purpose?* What claims does the author make about the state of the literature(s) at the time the review was written? What are the major debates and what are they about? How does it treat the major topics or concepts it identifies?

    b. *What does the author call for as an intervention into this broad literature's conversations?* What further or corrective work does the author call for? What is your assessment of the kinds of work needed to broaden or further develop disciplinary work in this literature category?

    c. *Are there concepts in the article that relate to your topic?* If so, which of these are relevant to your project, and how? If the review reflects a concept-fueled debate, such as whether to shift from studying water as a social element to studying it as an economic commodity, how is this debate relevant to your project?

    d. *Become multidimensionally minded.* How is your project connected, directly or indirectly, to this key literature? Or does your project have a different conceptual angle that points it to an expansion of this literature's key concerns or goes beyond this literature? If so, what is it? Freewriting answers to these questions can help surface your intuition.

Your answers to these questions help you zero in on how your project can respond to certain persisting, emerging, or reemerging disciplinary problems. Your answers can also help you to uncover obvious or intuitive concepts important to your project (and ultimately to proposal writing).

Moreover, answering these questions helps you figure out why you're doing the project in the first place. A project can't be fully justified because "no one's looked at this process before" or "nobody has studied these people" or "it's just interesting." Instead, a compelling project is grounded in a strong understanding of what would be unavailable to a creative or scholarly conversation, and to social and political knowledge at large, if the project doesn't happen.

Engaging with individual literatures in a clear, step-by-step way will help you make a case for the significance of your research focus and your design. With further reading, your understanding of its overall significance will be increasingly refined—something we help you do in Module 7.

Consider pausing and taking a twenty-four-hour break before moving on to the next exercise. Again, pay attention to how the body-mind-being is responding to your work. Ask what care is needed to cultivate or attend to what's arising.

EXERCISE 2C **Include Literature-Based "Other Concepts"**

While you were reading your literature review articles or chapters, you likely noticed that they have section titles and that there are broad disciplinary topics and ideas peppered throughout those sections. Authors highlight certain concepts to show their importance for scholars working in this literature domain, as well as to gesture to new topics and ideas toward which they should turn their attention. As we mentioned earlier, no doubt your research imaginary contains some of these literature-grounded concepts, in part because you've probably been reading already.

Now it's time for a little project expansion: adding key concepts from your literature searches. This exercise tests the relationship between your project and its key literatures. It requires you to consider how your project, when configured by lists of concepts, is in demonstrable dialogue with a community of knowledge makers. Do the following steps.

1. *Use your literature review work to identify three to seven additional key concepts vital for your project.* You probably saw several empirical or theoretical concepts that scholars are working with or questioning. These would be concepts that scholars are interested in or which they identified as lacking in current studies. Or, maybe you intuitively wrote about in your imaginary but, until now, you didn't think of them as directly relevant to the literatures you chose or relevant to a new literature. Now's the time to add them to your project. Be judicious: you can't study

everything! But be open to the wonderful process of oversight and opportunity.

Now you will need to find out where these new concepts belong. Use the prompts below to update your Key Concepts table, which will contain a new category, which we explain below.

2. *Expand and/or revise your 4Ps*. If the new concepts that you identified are relevant to the 4Ps, add them there. Some of the concepts you identified might illuminate the need for the addition of new subject groups (people?) or places (areas or spaces?) or processes. For example, a review of the anthropology of markets might suggest that computer programmers are increasingly important interlocutors that your project might need to include.

3. *Expand and/or revise your broad-context concepts*. If some of the new concepts you identified are broad-context terms that situate your ethnographic project in a broader social or spatial frame (for example, you found out that the anthropology of markets requires you to situate your project with the context of global "fin-tech," or financial technology), place such contextual concepts in the Broad Contexts box.

4. *Add other concepts*. We define "other concepts" as those that don't fit neatly into your project's empirical categories or broad contexts. Most of the time, these are highly theoretical terms found in disciplinary or social discourses that gesture to broad ideas or perceptions. For example, "constitutional law" or "intellectual property" or "jurisdictions" may be important and debated concepts that you found in legal anthropology literature reviews. And yet, it isn't a concept that you previously identified as a key empirical or broad context concept in your project's social worlds. So, you know you need to include it in order to be a part of disciplinary or social conversations about topics such as yours. You can populate this space with concepts from other sources as you take your multidimensional work further. See Rahman's work in Example 2.5. Bold terms are derived from literature searches.

EXAMPLE 2.5 Rahman's Key Concepts (Updated)

4Ps	*People*: real estate agents, plot owners, kachi abadi residents, developers, stamp faroshi, patwaris, investors *Parts (things)*: **liquid capital**, real estate deeds, paper files, digital files, transfer forms, WhatsApp communication platform *Processes*: real estate trading, Islamic investment, **regulation, formal and informal markets,** rizq, digitizing plots, urban planning *Places*: plots, Lahore, kachi abadis, mohallas, cities, informal settlements, city bureaucracies, family property
Broad Contexts	Global real estate trading, the Pakistani state, **Pakistani historical relations**, war on terror
Other Concepts	**Financialization, post-9/11 imperialism, governance, kinship relations**

Using his table as inspiration, create an updated Key Concepts table in Table 2.3.

7. *Contract the expansion and reduce to fifty-five or fewer concepts.* This upper-limit number of fifty-five concepts is an evidence-based suggestion for managing both congruence and overwhelm. Take a moment to make sure that you haven't expanded your table in an unwieldy way. Check your concepts and make sure they are all crucial to the project. Assess whether there are some you can consolidate or remove. You have to maintain project congruence, which impacts project feasibility (do you know what you're doing and can you do it?) and significance (what is this project's core value and why is it important?).

These literature engagement exercises are a lot of work: they may have taken a whole week! At this point, slow down and take a twenty-four-hour

TABLE 2.3 Key Concepts

4Ps	People: Parts (things): Processes: Places:
Broad Contexts	
Other Concepts	

break and attend to whatever needs are arising. After you've rested, do the next exercise.

EXERCISE 2D **Create and Listen to the Literatures Diagram**

You will create a literatures diagram for your project.

1. *Draw the diagram.* Referring to the literatures diagram and your work above, draw by hand or create digitally a three-part Venn diagram. Refer to Figure 2.1 as an example and fill in your literatures.

2. *Sit with the diagram and reflect.* Put the diagram up on a wall or visible surface and sit with it for a day. Afterward, write a re-flection about the adequacy or limitations of your literature se-lections, addressing the following prompts. You can lead with language like "I'm not sure but...." Or you can free-write, draw, make lists, or brainstorm in any way that feels right to you.

    a. *Listen for project congruence.* Describe how each of your literature selections, put in open relation with one an-other, support or make clear your project's overall schol-arly purpose and intention. For example, in a project about the changing political and medical circumstance of abortion in the US Virgin Islands, you could write, for example, "Putting the medical anthropology literature together with the anthropology of the US Virgin Islands will help orient me to where my work stands in relation to the current thinking on the relationship between US imperialism and reproductive health."

b.   *Listen for limitations.* Write about how the literatures you chose may *not feel like* they are sufficient for situating the project's purpose and intention fully. When we say "feel like," we mean in terms of your goals as a scholar or intent to create social transformation. Continuing with the Virgin Islands example, you might write "The medical anthropology literature currently focuses almost exclusively on biological processes, not on emergent liberatory or countermedical processes of well-being." You don't need to have an answer for what literature this points you to; simply write about it and see what comes up.

c.   *Identify potential project significance.* Describe how your project will connect with and contribute to or even transform each literature in significant ways. As you do this, write about how the literatures you chose may contribute to justifying the project's significance, either personally or in scholarly terms. Simply write about this and see what comes up.

d.   *Listen for other literatures.* Listen to your intuition about the literatures you've chosen. You might feel a little uncertain about them but, through reflection, have a stable overall sense of the current or eventual rightness or coming-togetherness of your choices. Begin to open to the possibility that there may be a literature that, in substitution or in addition, can multidimension your project in an innovative way. You'll return to this question in Module 4. For now, just resting with that possibility is enough.

Now take a moment to celebrate! You've identified three key literatures (for now) for your project, and you've reviewed the topical and conceptual milieus in which your project will sit. Consider taking another twenty-four-hour break before moving on to the next exercise.

**EXERCISE 2E Make a Preliminary Reading
List with Timelines**

This exercise simply helps you translate your literature review work into a
reading list that includes ten to twenty works that you should become familiar with. We consider this a launching exercise that you can continue at your
own pace in relation to the upcoming concept work. You may choose to do
this individually or share it: determine this process with your research design
community. If your group members have research interests or specific literature categories in common, consider using a shared file in citation management software to create a repository of useful sources. Even if your research
interests do not overlap, you may want to create a shared reading schedule
to help you maintain momentum on this task. Whether you create an individual or shared list and schedule, this process is a necessary step in the development of your relationship to literatures and community conversations.

To begin your reading list, consider the following prompts:

1. *Make an initial list of important works.* While reading your ar-
    ticles or chapters, you no doubt found authors referring to (1)
    canonical works, (2) broadly read newer work, and (3) emerging
    work that you haven't read. Some of these will be books, some
    will be articles. List at least two works in each category, for a
    minimum of six readings.

2. *Expand your list.* Next, review the bibliographic titles again, with
    an eye for titles that index concepts you identified in the earlier
    exercises. Select at least four other articles that pertain to key
    concepts. For example, if you were doing Rahman's project and
    focusing on economic anthropology, you would also need to
    zoom in on topical subsets, like *real estate markets* or *financial-
    ization*, which are vitally germane to the project.

3. *Create a reading plan.* Establish a reading plan to obtain a solid
    grounding within your key literature. Even if you are already fa-
    miliar with one or more of your literature categories, returning
    to key texts themselves can be surprising and generative. Make a
    reading schedule and begin to fill it in as you complete the mod-
    ules. Think about how you will strategize reading and organizing
    your notes and analytical engagements with your bodies of lit-

erature. Will you use a handwritten notebook? A Word document or Excel spreadsheet? Data management software? Choose something to organize your reading plan.

4. *Review your plan.* Make a note in your schedule about any areas where your understanding of your key literature is still lacking. Do you need to read more review articles or chapters? For example, have you located some key ethnographic studies, but you're not sure about what sorts of literatures you need to engage at the more theoretical levels of the project? Or something else? If there is nonalignment or other fuzziness, describe what that is and what you need to do to rectify it.

5. *Think about future reading.* Keep a running list of other citations you might want to read, and organize them according to the literatures that they belong to. Begin to note emerging patterns and concentrations. Is one obvious literature beginning to be clear? Is there a nonobvious literature on the horizon?

EXERCISE 2F Research Project Grid 2

Remember that each module will end with your reflections and additions to the Research Project Grid table, which gives you a bird's-eye view of the concept work you are doing throughout the modules. Each grid category will be explained with each upcoming module. We recommend that you save and date versions of your grid as you move through this process. You may want to revisit older versions in order to reflect on your past thinking and decisions.

As a reminder, the grid has several purposes:

– It's an assessment mechanism, which allows you to check for project congruence.

– It's a portable quick view of your research project, which provides a clear focal point for work with collaborators, advisers, and funders.

– It's an essential framework, which allows you to build congruent and integrated project proposal descriptions and narratives.

You can now begin filling in the Key Literatures row (upper half) of the grid! Place in this row the three literatures that you identified in this module. You will continue to add or adjust literatures, especially in Module 4. All of this work will help you construct a future research proposal literature review in a coherent, connected way.

Collective Concept Workspace 2

This module can be challenging, especially if you are just starting to figure out and examine your literature categories. If you are at the beginning, understand that it will take time to develop literatures. Moreover, it takes time to cultivate your own personal relationship to literatures. Think of it as a living connection that is always in motion, not something to be decided on and done with. That is, connecting to literatures puts you in direct relationship with a community of inquirers that is always present. Whether you are new to your literatures or you have already been working with them, allow yourself to settle into clarity, confusion, or simply not-knowing—whatever the experience may be at this point. Cultivating this sense is the most important part of the individual concept work and the group work. Keep in mind that congruence (or lack thereof) across these spaces (individual and group) is a good indicator of how the process is unfolding. Processing in groups helps to identify holes and can push you toward breakthroughs. If you continue to have blank spots, you can create a running list of issues to check in future modules. Trust that any questions or gaps will eventually reveal themselves in the concept-work and group processes.

In your group meeting, consider going over your agreements first before following the prompts and questions:

1. Read everyone's research imaginary and literature exercises for this module. In addition to any established feedback practices that your group has, try to identify what feels congruent, fuzzy, missing, and important for each.

2. Which exercises were difficult? Which were easy? Talk through and explain why. If any were difficult, it can be a sign that there is something fuzzy or noncongruent about the project conceptualization at this stage. Allow the difficulty to be an indication (and nothing more) that additional time and space are

needed for literature investigation alongside project conceptualization. Eventually this tacking back and forth between the literature and design will materialize deeper insights. On the other hand, if any of the exercises felt *too easy*, is there anything you're skipping over?

3. Discuss the following:

    a. Which three literatures did you choose and why?

    b. Which key concepts did you choose and why?

    c. How do the literatures and key concepts directly connect to the research imaginary?

4. If you are not feeling congruence between the research imaginary, the three literatures, and the key concepts, then work with your group to identify which areas need further concept work. You may need to take these actions (either during the discussion or afterward):

    a. Talk through the key arguments and debates in the review articles you selected and compare those to your own research project.

    b. Redo some of the exercises together. For example, you may need to talk about how you see your project contributing to the literatures. Or you may need to talk through your key concepts list.

    c. Rewrite your research imaginary. Incorporate any relevant feedback you've received, and that will additionally help make the literature and the key concepts clear.

MODULE 3

Map Concepts

You've elaborated your ethnographic research intentions and engaged the broad world of literatures. Those project expansion activities yielded your Key Concepts table. Now it's time to represent those concepts in a nonlinear way that will help you begin exploring your project's conceptual connectivity possibilities and, therefore, its emerging multidimensionality. You will do this via a one-page concept map. The concept map is a standard representational technique in many forms of research, creative work, and program planning.[1] Learning this skill will not only help you design an ethnographic project but also attune you to a mode of representation that is useful in many arenas. Concept maps can also act as foci for group attention, inviting discussion, and sparking collaboration.

MODULE PURPOSE
Create a one-page map of your project's key concepts.

Your map will also be a diagram of conceptual relationships. It will sketch out your project's intellectual field and scope. The map can also illuminate a deeper understanding of how your project relates to otherwise movements and possibilities. Building this kind of concept map will allow you to see your project as a multidimensionally imaginative form with many potentials. When you mobilize it again in Module 4, it will help you consider how its conceptual framework can reveal one or two innovative multidimensional concept combinations that help it stand out in relation to other research projects and literatures.

Concept-map making, as we coach you to do it here, cultivates project listening in a form that is different from structured writing or making lists. It is a "drawn" form that doesn't rely on normative scholarly forms of sequencing and ordering. Creating a form like this may be new to you. But even if it is already a familiar mode of representation, making a drawing of your project requires a different kind of creative openness than writing. Throughout the exercises, it's important to follow the prompts to pause and to build in your own pauses. Pauses will allow you to play with and listen to new connections between project concepts that arise when you free them from grids and writings. Remember that concept mapping doesn't have to be a lone activity; it can be done alongside or together with others. It's often quite fun!

To create your first concept map, you'll draw on all the work you have thus far generated, so gather it all together. As you embark on the map-making process in the first two exercises, you'll no doubt think of new concepts you could add. However, go slowly and be judicious. There will be opportunities to modify your map, but we want you to keep to a maximum of fifty-five concepts. As you work further to evaluate the concepts you map in the next exercise, you may need to prune, reshape, or completely reenvision the map. The final exercise in this module will help you make some of these revisions, which will set you up to define your multidimensional conceptual innovation (Module 4) and then write a research description (Module 5). These iterative moves are necessary processes and will inform the rest of your concept work.

Exercises

**EXERCISE 3A Prepare to Work Graphically
and Archive Your Creations**

Concept mapping requires your pictorial and intuitive sensibilities to be open
and relaxed. It's best to begin this process by making sure you're in a conducive state of mind: rested, open-minded, curious. Sometimes, creative picto-
rialization is difficult for people who feel anxious about drawing or graphic
representation. If you feel that way, be sure to acknowledge it and cultivate
compassion for those experiences rather than trying to ignore them. Know
that you will be able to choose the medium that is most comfortable and
accessible to you.

Before you go further, spend some time connecting with your aesthetic
sensibilities and getting inspired. We offer these prompts, in a suggested order,
to help you explore representational possibilities.

1. *Appreciate nonnarrative forms.* Take a moment to reacquaint
    yourself with images or sounds or other aesthetic forms from
    your favorite artists or artistic works. Think or free-write about
    how and why these forms are meaningful and inspire you. Here
    are some prompts:

    a. What do they spark in your mind and heart? How do they
        do this?

    b. What kinds of messages and effects do nonverbal, non-
        narrative forms convey (ideas, intentions, feelings, argu-
        ments, values, moods, perspectives, etc.)?

2. *Look at concept maps.* Spend some time online searching for
    "concept maps" and "brainstorming images." The results will fa-
    miliarize you with some of the shapes and configurations that
    people use (tree-like, rhizome-like, bubble maps, among others).
    Some of the standard shapes include branches, bubbles, and
    webs. Notice the personalized approaches to the use of digital

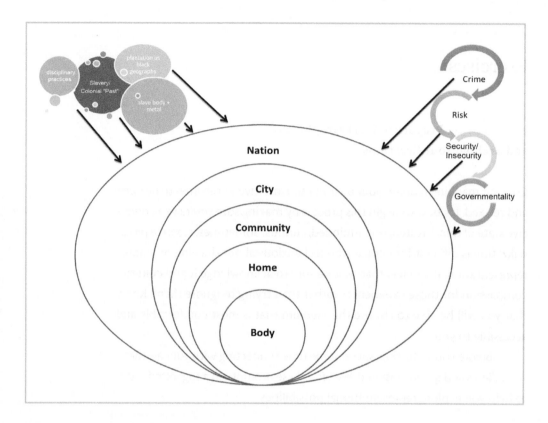

FIGURE 3.1 McKinson's Concept Map

shape and handwork between two students' early draft "bubble maps" in Figures 3.1 and 3.2. Also appreciate how, in both, you can detect the key ideas and connections that animate each project. You are producing a similarly personal and process-oriented graphic expression.

3.   *Free-write about the map form examples.* As you explore mind maps online, ask yourself these questions and free-write about which approaches appeal to you and why. Here are some prompts:

    a.   Why do you prefer certain map shapes over others?

    b.   What is it about this form of representation that inspires you or daunts you?

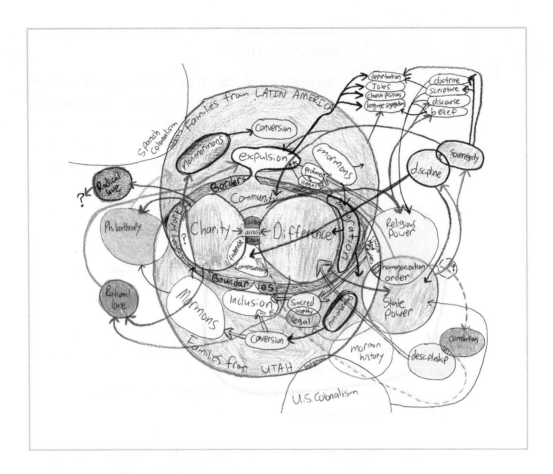

FIGURE 3.2 Palmer's Concept Map

4.   *Think about mapping media.* After relaxing your mind with pictorializations that inspire you, you'll need to choose a diagramming medium that you feel comfortable with. All are acceptable! Hand drawing, cutting and pasting, and using a digital application all take you to the same place. The only requirement is that you have a way to move concepts around and experiment with composition. Here are digital or handwork mapping options:

–   *Drawing a map by hand.* If you go analogue and draw by hand, consider drafting your map using a sticky note for each of your terms or phrases so that you can easily move them around. This will make it easier to revise your map. Hand drawing can also take place in an artist's sketchbook, where you can keep track of concept-map iterations over time.

— *Drawing a digital map.* You may wish to use a concept-mapping program. We suggest that you try some hand drawing first just to see where it takes you, and then choose a digital program if that feels right. There are many free concept-mapping or mind-mapping programs. These offer the advantages of preformatted and changeable shapes and designs, but those could constrain your initial attempts to find a shape that is meaningful to you. You can also use software that is already familiar to you, like Microsoft PowerPoint, Miro, or Adobe InDesign, among others.

By the end of this module, you'll want to land on a final concept map that pleases you and can be shared, whether it is hand-drawn, digitally generated, or some combination of the two.

Here are some key organizational strategies that will help you with all your iterative concept-map work:

5. *Archive your drafts.* Make sure you have a way to archive and date multiple iterations of the concept map and corresponding reflections so that you can revisit your conceptual thinking over time. If you choose a free digital program, will you lose your designs over time? If you work on paper, do you have digital backups in case the originals aren't accessible?

6. *Recognize the value of returning to old drafts.* It is important to save your work because if you hit a wall later on, you can return to previous maps to get an overall graphic view to refresh your perspective. You can also evaluate whether there are any important project concepts—such as those that provide simple clarity about who or what is at the core of the project—that got dropped along the way but need to be picked up again. Concepts often get dropped because they are not fully formed or clear yet, even if they have intuitive potential. In the long course of research design, many researchers spiral out into conceptual complexity. Or, conversely, they go down narrow ideational

rabbit holes but eventually return to original design work to re-calibrate themselves. Returning to old maps can help to sharpen fuzzy thinking and reengage intuitive directions.

EXERCISE 3B **Make a Concept Map with Connections**

It's time to map! In Module 2, you took opportunities as needed to revise your research imaginary and project concepts, which culminated in your full Key Concepts table (completed in Exercise 2c). This is the time to slow down and really look at the table: Does it have the key concepts that are defining your project and its multidimensional possibilities? Revise as necessary.

Your current Key Concepts table will be the basis for your concept map. For simplicity's sake, we will refer to all the terms or phrases in it as *concepts*.

1. *Manage your concepts.* We recommended in Module 2 that you have fifty-five concepts or fewer, but you may have more. If you have concepts you do not wish to let go of, devise a way to keep track of them. These may come in handy during the mapping process, as a source of more effective concepts you can swap in.

2. *Arrange your concepts on a digital or analog page to make the beginnings of a map.* To begin mapping, don't worry so much about the *form* that the concept map will take. If you're using mapping software or have decided on a form, you can work with branches, bubbles, or webs or a hybrid of these shapes. But if you're not following a format, just begin putting your concepts down into the analog or digital space. Use techniques such as sticky notes or movable digital shapes to help keep your concepts easy to move around. Know that you may change design strategies as you move along in the process.

3. *Convey information about your concepts via color, relative shape or size, and arrangement.* Typically, people nest concepts within one or many kinds of shapes, give them different colors, and vary their size and proximity to one another. For example, you may use different colored and sized shapes for the 4Ps, the Broad Contexts, and Other Concepts. You may also use different fonts, colors, or sizes to convey differences in the relative empirical,

theoretical, political, or social "weights" of your different concepts. Arranging your concepts mindfully in terms of nearness or distance from one another, and relative distance between center and edge placement, allows you to display a variety of qualities. For example, you may cluster concepts to represent social subworlds of your project or different broad processes. Or you may match certain 4P terms to particular broad-context terms. Experiment with various forms and arrangements, but make notes on what they mean so that you can narrate this in your reflection.

4. *Convey conceptual connections.* At this point, it is perfectly fine to focus on shapes and groupings, but it is also important that your map conveys key conceptual connections, which is an important first step in multidimensioning. We will prompt you to narrate those relationships, but at this point you can begin to render them graphically. Most of the time, concept maps use shapes, lines, or arrows to represent connections between concepts. These lines of connection can be creatively differentiated. For example, some research designers use whole lines to represent strong connections and dotted lines for weaker or less apparent but nonetheless intuitively clear connections. You might use connective elements to pair two concepts or to create constellations of multiple concepts. In addition, some maps link multiple concepts in forms such as branches, webs, or hubs and spokes. Some use varying colors to indicate different dimensions of the project, such as flows of energy, particular key processes, or relationships.

Don't feel pressured to create connections everywhere, or to know all your links. If you feel a connection but are not sure why you do, consider sketching lines and putting question marks within them. Those marks will be placeholders for further iteration.

The goal of mapping is to convey, to yourself and others, important information about the state of your project's composition and congruence. Remind yourself that this is an ongoing and iterative process: the map that you create represents what you know right now about your project's components and connections. With further research and concept work, your map will change to reflect what you learn along the way.

Finally, don't be worried about whether your map "looks right" or "looks good." What matters now is that you have completed two crucial project "whole" forms that can be related to one another: a narrative research imaginary and a graphic concept map. You will use these two forms to develop and refine the multidimensional object that connects it all. This will help you to contract the project when it balloons, to reimagine connections, and to communicate your conceptual framework to others.

It's time to look at the map's conceptual landscape in terms of its ethnographic congruence. Make sure that you evaluate whether the terms you are using are well chosen and are relatable, even if at a conceptual distance. The reason you do this assessment now, after you've done the map, is that you can perceive these concepts differently than when they are sequenced in a narrative or categorically listed in a table. In a concept map, they are put into dynamic relationships with each other, making it possible to think about how valid, accurate, and relevant they are to your project's overall congruence.

EXERCISE 3C Assess Your Map's Ethnographic Specificity

This next step of the map making process is vital. It will entail asking you to question your most disciplinarily general terms—such as neoliberalism, nationalism, development, capitalism, and so forth. We have found that students can over-rely on these disciplinary terms and use them as placeholders. Such placeholders can be helpful for signaling broader processes. However, they are so commonly used and so ubiquitous in research that they simply obscure what's happening in your project. We have seen lots of researchers become deeply attached to these hegemonic terms, which leads to frustrations when a project cannot name what is specific about the research problem, processes, and contexts. In addition, supporters and funders cannot determine what is real and compelling about your project.

In this exercise, we help you avoid this common and sometimes catastrophic design pitfalls. We show you how placeholder terms can be made more ethnographically powerful as research design concepts.

The first step is to understand the basic structure of an ethnographic project, which is essentially a conceptual hybrid. It includes concepts that are mainly empirical, those that are mainly theoretical, and those that are both. What we mean by this is that there are concepts that are recognizable to those whose worlds we conduct research in, concepts that are meaningful to others who use research to produce generalizable knowledge, and concepts that are both recognizable and meaningful in both senses.

In traditional linguistic anthropological parlance, these conceptual categories are often referred to as *emic* (meaningful to the group being studied) and *etic* (meaningful to those who are studying them). This can be useful but can also presuppose de facto hierarchical separations of meanings and stakes between researchers and interlocutors. We do not automatically support that model. But we also recognize that it is useful to recognize terms meaningful within fieldwork engagements and terms that reflect the creative and analytic process of making new research works and concepts. Table 3.1 provides a useful way to distinguish the two kinds of concepts.

TABLE 3.1 Empirical vs. Theoretical Concepts

Empirical Concepts	Terms for specific people, processes, places, things, and ideas that are conversationally and experientially meaningful to your interlocutors and you; concepts that name beings and things that you will directly engage when you do fieldwork.
Theoretical Concepts	Generalizable terms and ideas that are important to research communities and the production of knowledge, insight, and liberation; often these are terms that you engage as you analyze, create, and disseminate new knowledge and representations.

The most effective concept maps are those that display the empirical and theoretical concepts that cohere the social and intellectual legitimacy of your project. Not every project has a perfect balance between the two kinds of concepts, and very often concepts can belong to both categories. Concepts can also occupy both categories, depending on the addition or subtraction of additional terms.

Let's explore an example. Based on your literature review, your project in Brazil will aim to rethink the theoretical concept of democracy. Therefore, you will include the unlabeled term *democracy* as a theoretical concept because it is relevant to the literatures you are trying to contribute to. However, you may also choose to include *Brazilian democracy* in your project as a key empirical concept because it is a specifically unique process that you will be observing or participating in. Therefore, the term "democracy" will show up in your Key Concepts table and concept map, in two forms: *democracy* and *Brazilian democracy*.

The goal for mapping is to identify general and specific concepts that your advisers, community, and supporters would recognize as important to attend to and theorize. Overall, you need to know why you're choosing to include certain kinds of concepts, why you've phrased them as you have, and to have a good sense of how they interrelate—in harmony and tension—to create your unique framework of inquiry.

Often, in the early phases of the research design process, a designer won't take time to assess the concepts they are using and why. One result of this is that they choose generic empirical concepts that might be better expressed in terms closer to those being used by interlocutors. Or they will choose a theoretical concept but exclude an empirical concept that should also appear on the map. Or which should be used instead of the theoretical concept. This may occur because the term being used turns out to be relevant to literatures or projects other than the researcher's own project. Take a moment to assess and revise your terms as necessary. Follow these steps:

1. *Identify your concepts.* Choose a highlighter pen, or your computer's highlighting function, to represent an empirical or theoretical concept. Circle or highlight each concept on your map to indicate whether it is primarily empirical or theoretical. If it is irreducibly both, highlight it with both colors.

2. *Clarify your terms.* Are there any concepts that you can't identify as empirical or theoretical? This usually means that there is a more accurate term to use. Refine your concepts if necessary and then highlight them.

In the next steps, you will assess the phrasing, appropriateness, and ethnographic usefulness of your empirical and theoretical concepts.

3. *Assess the specificity of your empirical concepts:* Could any of your empirical concepts be better expressed in terms closer to

those being used by interlocutors? Would that bring more ethnographic precision, and therefore theoretical integrity, to the project's conceptual framework?

Consider this example. If you are going to investigate a process that is unfolding in Mexican agricultural spaces, you may use the concept of farmlands to represent your fieldsites. But, upon reflection, perhaps after seeing farmlands next to the concepts agrarian reform or NAFTA (the North American Free Trade Agreement) on your map, you may realize that you are interested specifically in *ejidos*, which are farming lots with a particular history. They also signal farming precarity in the context of neoliberal legislation and privatization in contemporary Mexico.

When you make a conceptual substitution like this, you are not just using a more ethnographically precise term. You are also replacing it with a concept that drives how a particular state structures relationships between persons, institutions, land, plants, and animals. Attention to this specific kind of Mexican farmland, the *ejido*, can impact how other concepts like property or land are understood anthropologically. This makes the new term more useful for ensuring the clarity and robustness of your ethnographic and theoretical aims.

Now take the following steps to assess your empirical terms. Do you need to modify or rephrase them, replace them, or make them less generalizable and more specific for your project's aims?

a.   *Assess concepts for social and cultural specificity.* See if there are any empirical concepts that could be modified in ways that more directly reflect the lived and situated experiences of the worlds you are engaging. Circle them.

b.   *Rephrase the language.* As indicated in the ejido example above, adjusting empirical concepts may entail redesignating people, beings, things, or processes in terms that are linguistically germane to your interlocutors and sites. Making this choice is also crucial for transforming the linguistic politics of ethnographic writing and theorizing. Look at the concepts you've circled and replace or rephrase them in ways that are more ethnographically grounded in the particularities of your project.

4. *Assess whether you need to change a disciplinarily theoretical concept into an more site-specific theoretical concept.* Are you using a theoretical concept that is meaningful to disciplinary discourses at the exclusion of a general concept that is vital to what is at stake for your interlocutors?

For example, if you are doing a project about the violent treatment of LGBTQI people in public spaces in Texas, you might have chosen *structural violence* and *spatial sovereignty* as general terms that represent this dimension of your project. However, as you look at *structural violence* on your concept map, you notice that it is sitting next to the term *public spaces*. As you think about the exact publics that you are studying, you realize that the term that your interlocutors use to define what they are seeking is *safe spaces*. This concept relates to theories of *spatial violence* and *spatial sovereignty* in anthropological literature, but it is also a concept that people are using to theorize their own experiences and to change spatial power and exclusion dynamics.

Your recognition could justify adding *safe space* to the map or using it as a substitute for the more disciplinarily theoretical term. Or you could add *safe space* to the map and make the disciplinary theoretical terms more empirically specific, such as reframing them as *Texas spatial violence* or *Texas spatial sovereignty*.

Now, use the following prompts to assess your own work to see if there's a reason and opportunity to be more theoretically and empirically specific with your concepts.

a. *Test for theory concept deletion or revision.* Look at your map and consider each of the theoretical concepts you've included. Ask yourself: Did you really need this broad disciplinary term in your project? If so, keep it. Or would your project conceptualization benefit from a more specifically empirical or theoretically situated term? If so, continue with these next steps.

b. *Assess whether theory concepts can be empiricized or transformed from disciplinary terms to more site-specific theoretical terms.* Use a different manual or digital color to circle any concepts on the map that you suspect might be better phrased as empirical concepts or as theoretical concepts more relevant to your interlocutors' experiences. Now let's test each circled concept for replacement or modification.

c.   *Test for replacement.* Is there an empirical concept than can be an ethnographically effective substitute for a theoretical concept? Address this by seeing if you can use a more site-specific version (*safe space*) of a theoretical concept (*spatial sovereignty*) and make that substitution. Take a moment after each substitution to see what happens to your map. Assess whether the substitution enhances the ethnographically particular nature of your project.

d.   *Test for modification.* Sometimes you can create a more effective theoretical concept for your project by adding an empirical term to a theoretical term or phrase. For the concepts you've circled, consider whether adding an empirical specifier (*Texas*) to those terms (*spatial violence*) creates a concept map that better reflects your theoretical aims. If so, make those changes.

Note how more ethnographically robust empirical and theoretical concepts provide you with a compelling way, in a proposal, to create a link between how you situate your project within literatures and how you describe your data-collection plan. We will examine the importance of such concepts more thoroughly in Modules 7 to 9.

At the end of this exercise, you should have a carefully considered and arranged concept map. If you feel the need to make any more adjustments, do so before completing the following reflections. Consider taking a twenty-four-hour break before proceeding.

EXERCISE 3D   **Listen to and Explain Your Map**

You will now listen to your completed and adjusted final concept map. The mapping process is about allowing ideas—whether initial or fully formed—to percolate in ways that prepare you for both fieldwork and theory-making.

Now that you have settled on a concept map that rings true for now, write a one-page reflective explanation. Answer the following questions, which will alert you, as with your research imaginary, to assumptions, persisting problems, and questions.

1. *Write a narrative explaining the map.* You can free-write, use voice memos, or try other methods. But be sure to create an edited version that you can share. Spend time explaining concepts, connections between particular concepts, and bigger constellations of concepts.

2. *Compare your map with your research imaginary.* What does this exercise tell you about your project that is different from what you wrote in the research imaginary? What feels missing or incomplete? What was newly included?

3. *Reflect on the relationship between your mapped concepts and your literatures.* Can you perceive how your map's conceptual connections will contribute to these literatures?

Keep this reflection in your files with your concept map. You may not have answers to these issues now, but you've articulated them and considered them as an act of project listening.

EXERCISE 3E **Research Project Grid 3**

Now that you have a concept map, you can go back and adjust your Key Concepts table, reflecting any rephrasing, substitutions, or additions to the table categories. After you update your table, turn to your Research Project Grid table and import the relevant changes to 4Ps, Broad Contexts, and Key Literatures (in the rows found in the upper half of the grid), if necessary.

Collective Concept Workspace 3

Even though this module invited you to contract the expansive work you've already done, it probably also revealed parts of the project that need more unearthing or clarification. This can lead to excitement, discouragement, and other emotions and reactions. Bring your concerns to the group and make space and time to process them. Know that you might have some break-throughs. Know also that it may simply just take some time for connections to reveal themselves. When things feel slow or out of reach, think of this process as meeting someone for the first time and cultivating a friendship over the long term. In other words, during times of challenge and struggle, see if you can befriend your project. Remember also that producing a shareable form like a concept map is, like writing, a process that is personal and, for some, laden with concerns about doing pictorial work. Be respectful of the unique approaches of your group members. Before starting the group work, consider revisiting your agreements in a way that feels right to you.

1. Look at everyone's concept maps and read their reflections. Each person can present a brief oral orientation to their work and what they are discovering. Discuss what was exciting, difficult, frustrating, easy, effortless, and so on while doing the exercises for this module.

2. Why did you choose to work with a particular map form and tools? How did it feel to represent your project graphically?

3. How did it feel to narrate, in broad language, the various concepts and concept constellations of the project (without using broad conceptual terms and placeholder language)? How well did that work? How might it have been difficult?

4. Talk about how and why you arranged the different project concepts in the way that you did. What connections did you make between concepts and why? What insights did these connections and arrangements reveal for the project?

5. Leave time for everyone to discuss what seems fuzzy or incomplete on maps. Also, discuss the breakthroughs you would like to have with your map and concept work generally.

Create Multidimensional Concept Combos

In this module, you will engage this handbook's central aim: designing multidimensional projects by combining key concepts in innovative ways. You will identify at least one fruitful concept combination, or concept combo, that establishes your project's particular scholarly contribution. Your concept combo brings three to four empirical and theoretical concepts together in a demonstrably creative way, showing that your project offers a *new framework of inquiry*. Because these concepts also connect your key literatures, the combo showcases the exciting intervention your project can make to those conversations. Both acts contribute to project tensegrity: holding concepts and literatures in productive tension.

MODULE PURPOSE
Locate a key concept combo that represents your project's multidimensional contribution to literatures

---

All in all, your concept combo opens a way for your work to uniquely contribute to and transform existing research worlds. To develop this combo, you will build on previous concept work: in particular, Module 2's literature work. This process introduces you to project iteration. That is, you will return to a previous concept-work stage to build multidimensionality and strengthen congruence. During this process, you will have a chance to confirm or revise your set of key literatures. This ensures that your project will engage the most relevant scholarly conversations.

This module also represents a project development phase shift: from project expansion to project contraction. In the first three modules, you narrated an imaginary, developed a broad set of key concepts, identified key literatures, and mapped your concepts. This module's concept-combo work *contracts* the project's diverse conceptscape into a small set of connected terms. The concept-combo terms work as a multidimensional anchor for the next phases of project contraction and expansion.

In any kind of mind-body training, learning to flex inward and outward builds your capacities. In multidimensional research design, such flexibility makes concept workflows and iterations feel satisfying and generative. Project development becomes less linear and mechanical and more akin to other multidirectional life flows like moving, growing, breathing, and relating.

Take a moment to reacquaint yourself with multidimension*ing* as a design process. *Multidimensioning* is the process of defining a project's conceptual combinations and using them to create congruently integrated project elements, from a research description to research questions. Your concept combos both reflect and initiate project congruence.

The concept-combo generation process is simple and illuminating. You will place all your key concepts into your literature Venn diagram in order to play with and perceive potential crosscutting combinatory magic. We find that this works well after Module 3 because you've developed a practice of moving concepts around and seeing how they relate to one another. In this case you're looking for a combo that is not necessarily common or commonsensical. But rather, one that you experientially and intuitively suspect will open up research dimensions that haven't yet been well examined by others.

You might have begun to sense potential multidimensional combos when you did your first round of work to combine literatures with each other. You may have sensed that there were project-relevant concepts in those individual literature categories that you could bring to structure new inquiries into changing social conditions and dynamics. Whether or not you saw them back then, you'll identify such promising concept combos now. You will confirm those with the best multidimensional potential by checking research databases to see what they reveal in terms of existing studies. In this process, you may discover that you can replace one of your literatures or add another one to your existing three. Doing so will support a cross-literature concept combo that catalyzes multidimensionality. We provide examples along the way so that you can see the genesis of conceptual multidimensionality in the projects we presented in the introduction.

You're looking for combos that have the potential to bring key literatures into new relationships. The most inspiring combos to work with are those that include a concept that *crosscuts* your key literatures and another that is *unique* and unshared among all three. Both situations are relevant to multidimensioning because they offer ways to build on existing scholarly conversations and to add difference to them. This dynamic is what validates an ethnographic project that must be—and wants to be—done.

Let's work through an example from a project you are familiar with so far: Tariq Rahman's. Typically we provide examples during or after exercises, but we want to walk you through the thought process before you begin, because it will help prime your concept-combining imagination.

If you remember, Rahman ended up with a set of three literatures: Pakistan studies, economic anthropology, and urban anthropology. Looking at his Key Concepts table (repeated here as Example 4.1), you can see a

EXAMPLE 4.1 Rahman's Key Concepts

4Ps	*People*: real estate agents, plot owners, kachi abadi residents, developers, stamp faroshi, patwaris, investors *Parts (things)*: liquid capital, real estate deeds, paper files, digital files, transfer forms, WhatsApp communication platform *Processes*: real estate trading, Islamic investment, regulation, formal and informal markets, rizq, digitizing plots, urban planning *Places*: plots, Lahore, kachi abadis, mohallas, cities, informal settlements, family property, city bureaucracies
Broad Contexts	Global real estate trading, the Pakistani state, Pakistani historical relations, war on terror
Other Concepts	Financialization, post-9/11 imperialism, governance, kinship relations

set of concepts shared between all three categories might be *land, finance,* and *governance.*

A key concept that seems to be only tangentially related to any of the literatures, and therefore would sit outside of their circles on a Venn diagram map, is *digital technology,* by which he means the kinds of technologies that make land transactions occur more rapidly in Lahore by bypassing the deed structure of the past. This process is transforming Lahore's land into financialized flows of global real estate capital. It is also formalizing the involvement of the Pakistani state and transnational governmental regulators in Pakistan's urban life.

This prompted Rahman to think about the ways technologies paradoxically secure and transform the grounded, regulated, and unregulated materiality of land markets. While he was at this midcourse stage of literature work, he wrote this concept-work reflection:

> I see my project as situated between two bodies of literature: economic anthropology and urban anthropology. Economic anthropology might be described as focusing on the cultural mediation of economic phenomena, such as markets, money, finance, gifts, labor, and so on. Urban anthropology explores how cities structure, and are structured by, broader anthropological objects of inquiry, including time, knowledge, racism, surveillance, politics, citizenship, violence, and much more. I am still struggling with identifying the third body of literature. My project took an interesting turn this week, which might lead me to delve into Technology Studies, and in particular work on material agency.

Rahman's statement that his concept work "took an interesting turn" reflects the moment he committed to paying ethnographic attention to the materially changing economic and political nature of the real estate "plot" in Pakistan as well as its global presence. He noticed that a plot itself seemed to be broadly understood by mohalla residents and government officials as having a technically facilitated fluidity in the world rather than being simply a local territorial site.

Before he began to conceive of a plot this way, Rahman was still situating his work within traditional vertical spatial scaling schemas, like local city to nation to globe. Then he began to see how a focus on plot digitization could link his literatures and open new dimensions of inquiry. Specifically, he opened the project to new questions about how plots had become something other than immobile real estate; and about to whose advantage and disadvantage this new mobility was. In the end, he didn't add the science and technology studies literature. But he recognized that digital technologies and new transnational legal and economic connections, facilitated by imperialism and financialization, were making investment in real estate plots less concrete and more fluidly mobile; this was an important, uniquely positioned concept that would make his project multidimensional.

To check his project's potential multidimensionality and scholarly contribution, Rahman could search on article abstracts or full texts using several crosscutting combos of three to four concepts that have empirical and broader theoretical or contextual concepts. When we represent this in the handbook, we use plus (+) symbols to indicate the Boolean structure often used in digital searches. The best combos, for Rahman, focused on the key dimensions of his project that are linked to his key literatures. For example, to use concepts that reflect his focus on *financialization* (which links to economic anthropology), his focus on *city land transformations* (which connects to urban anthropology), and his focus on *South Asia* and *Pakistan*. Limiting the search combos to three to four concepts keeps the search open enough but limited enough to yield meaningful results.

The numbers of articles and books one finds using concept combos can vary, but the point is to figure out where your project sits within and across research worlds. As he pushed his project further and did more literature searches, Rahman landed on concept combos that opened new directions for him and that were obvious ways to contribute to scholarly conversations. The most promising were

- *Pakistan + real estate + financialization*
- *land + South Asia + digital technologies* and
- *cities + Islamic investment + governance.*

Through those searches, he discovered that scholarship that addresses land transactions in Pakistan was focused on internal, national-level economies and governance structures. He also found that the scholarship did *not* focus on the relationship between those processes and the entry of Pakistani land into a global real estate market or on the particular legal, economic, and political modalities making that possible.

In the example we provided, all the concept combos revealed that there was little work being done at the time on the ways in which urban spaces in South Asia were becoming abstractly financialized within global flows of capital and investment. Rahman's next phases of concept work would take him further into exploring the significance of his project's focus on Lahore real estate transactions and urban family strategies of money-making; these are being shaped by Pakistani diasporic business worlds and racial capital configurations like the war on terror. As you already know from our previous discussions of this work, playing with concept combos like this allowed Rahman to discover his multidimensional object (MO)—*liquid land*. This MO represented the conceptual coherence and scope of the project because it points to the transformation of land into financially, culturally, technically, and politically fluid space; these dynamics take place in the context of global investment markets impacted by world banking regimes and US imperialism.

Finding this kind of conceptual clarity in your project is an ongoing process. You will continue to fine-tune this process as you move ahead. But this phase will ground you in the techniques and spirit of multidimensioning, which is about opening portals into new ways of knowing shared social experiences.

After this module, you will use your concept combo to compose a research description (Module 5) and to perceive the MO that holds the whole framework together (Module 6). The MO you land on may or may not contain terms from your multidimensional concept combo. But having that combo on hand will help ensure that your MO connects your whole set of key concepts with your literature work and your main research question. After this, you'll reflect on your combos to pose an overarching research question (Module 7). This will allow you to develop data sets and data-collection questions (Modules 8 and 9) that define the specific work you'll do in the field.

Exercises

EXERCISE 4A Place Key Concepts within Literatures

In this exercise, you will use your literature Venn diagram to place each of your key concepts within the key literatures. This allows the Venn diagram you created in Module 2 to become a multidimensional portal. By "portal," we mean that it will create a space for moving between and across literature categories in ways that open new dimensions of inquiry. In some ways, this activity upholds hegemonic Western categorizations of knowledge. But it also opens space to reorganize and transform the content of those literatures as well as to undermine hard boundaries between them.

You're ready to move all your key concepts into your Venn diagram. We recommend doing that on paper, because you can create a diagram large enough to play with the placement of individual key concepts in a free and embodied way. Use an erasable pencil or pen and sticky notes that can be moved around easily. However, if you wish, you can also use a word-processing or graphics program instead—but do so on a screen that allows you to perceive the concepts clearly.

First, see the results of this process via Kimberley D. McKinson's (Examples 4.2 and 4.4) and Jason Palmer's (Examples 4.3 and 4.5) projects. We provide their Key Concepts tables and Venn diagrams that represented the concept work relevant to their projects.

The concepts in these tables and diagrams are phrased in the ways we recommend: as clear phrases, of one to three words each, that clearly encapsulate the concepts relevant to the project. This is the time to make sure your table is clear and concise. You may have checked this in Module 2, but now it's even more important.

Your key concepts must also be good search terms, which means they must represent clearly defined aspects of your project. For example, take the example of a concept phrased as *citizenship documents with multiple bureaucratic purposes*. You realize that you and other scholars are interested in citizenship documents in general. However, your current phrase won't allow you to assess whether other scholars have either looked at citizenship documents or overlooked them in relation to other concepts, such as *migration, religion*, or other key research terms. Consider whether such a phrase needs to be better framed, or broken apart, such as to reduce this phrase into *citizenship documents*; or perhaps to break it into *citizenship documents* and *bureaucratic procedures*. In addition, students sometimes have multiple theoretical phrases in a Key Concepts table that won't yield a hit in searches, such as *oppressive mechanisms of the carceral state*. This is more like a sentence than a project concept. It can be broken into component parts that reflect the project's 4P focuses, contexts, or theories, such as *carceral state, disenfranchisement*, and *redlining*.

EXAMPLE 4.2 McKinson's Key Concepts

4Ps	*People*: artisans; residents; architects; construction workers; police; security employees; Neighborhood Watch; corporate stakeholders; foreign funders; plantation museum staff
	Parts (things): the body, metal security structures, electronic security devices, recycled metals, home security
	Processes: criminality/victimhood, metalworking, surveillance, policing, securitization, citizenship, dwelling, aesthetics, discipline and enslavement
	Places: Caribbean, Jamaica, residences, artisanal workshops, construction sites, neighborhoods, security company offices, Neighborhood Watch meetings, museums, crime prevention workshops
Broad Contexts	Postcolonial states, security regimes, colonialism in transition, global Blackness
Other Concepts	Slavery, Black geographies, carcerality, citizenship

True to the spirit of this exercise, McKinson's conceptual placements reflect what she perceived about the state of the anthropology of Jamaica, Black studies, and material culture. For example, it is clear how the concepts of colonialism, slavery, and security are present in all three literatures (see Figure 4.1). And there are also some dually shared concepts such as criminality and discipline. She was familiar with the treatment of colonial-era metal objects meant for enclosure, enslavement, and murder following her historical work on the anthropology of Jamaica and in Black studies (so she placed these in overlap spaces). During preliminary fieldwork in Kingston, she was drawn to the many elegantly crafted security gates and fences to be found in certain neighborhoods. She had been repeatedly struck by the importance of artisanal metalwork to the experiences of people embedded in the historical and contemporary insecurities of living in and traversing neighborhoods across the city. But she wasn't sure she had seen any mention of *security metalwork* or *home security devices* in any of the literatures, placing these uniquely

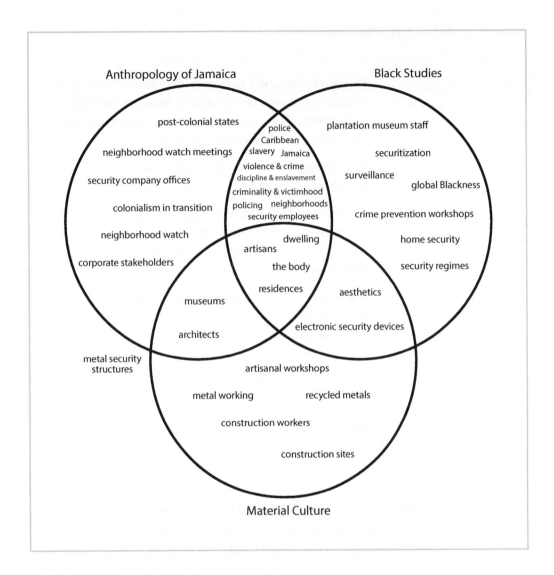

FIGURE 4.1 McKinson's Concept Placements

outside the literatures. These are the kinds of intuitive observations that can inform a designer's recognition of project multidimensionality, and the ways it can connect literatures.

Jason Palmer's work reflects a similar realization of concepts that seem unique to some literatures and those that sit at intersections. During his preliminary fieldwork, Palmer interacted with Peruvian Latter-day Saints (LDS) church members (Mormons) who were building a new temple in their

EXAMPLE 4.3 Palmer's Key Concepts

4Ps	*People*: Peruvian Latter-day Saints (Mormons), Utah (US) LDS citizens, noncitizens, church officials, Peruvians with LDS relatives, intermarried spouses, missionaries, Peruvian government officials, Utah government officials
	Parts (things): immigration documents, souls, borders, temple records, family genealogies
	Processes: citizenship, racism, church membership, religious practices, migration, love, conversion, pilgrimage, lineage and religious community practices, pioneer discourse, "testimonies" (ritualized truth claiming), mobility, il/legality
	Places: Peru, Utah, homes, wards, churches, general conferences, temples, missionary training centers, immigration offices, embassies, job sites, government buildings
Broad Contexts	US colonialism, LDS global and cosmos-scaled church, nation-states, religious migration, Peruvian religious practices
Other Concepts	Religious power, religious discourse, state power

community and also struggling to find ways to travel to or migrate to the US state of Utah. He heard people express their contradictory existences in the spiritually overlapping but painfully and often violently segregated worlds of Peruvian and white US LDS community membership. Searches on concepts from his concept map (see Module 3 for one version of his map) revealed that *membership* is a concept that is more pertinent to the anthropology of religion than to political and legal anthropology, which is a literature category more focused on citizenship. In addition, *love* and *the soul* were present in the anthropology of religion and anthropology of Peru literatures. However, they were not, as far as he could tell, in migration studies. With his diagram (see Figure 4.2), it's possible to see how Palmer's key concepts could be conjoined to make a cross-disciplinary contribution.

The first step of this exercise is to build on your previous literature searching to represent the specialized or generalized status of your concepts in key literature conversations. Follow these prompts:

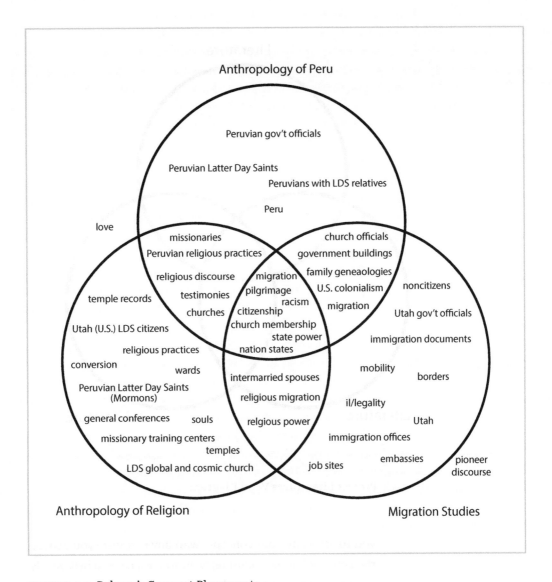

FIGURE 4.2 Palmer's Concept Placements

1. *Create a large version of your Venn diagram.* Make a Venn diagram with your literatures labeled on the sides with circles large enough to write terms within (see Figure 4.3 as an example).

2. *Prepare to place concepts within your diagram.* Have your Key Concepts table on hand. The spirit of this process is to establish a preliminary sense of where your concepts actually appear in literatures and where they don't. Therefore, it's important for

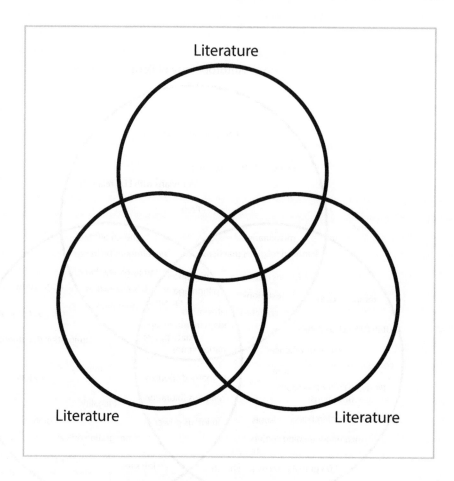

Literature

Literature

Literature

FIGURE 4.3 Project Literatures Venn Diagram

you to associate your concepts with any literature you believe the concept has been *most* relevant to—even if you believe it is a concept that deserves to be relevant to another of your literatures. Here are prompts for placement decisions:

a. *Does the concept seem to fit within only one of the literatures?* A concept may fit squarely within a literature (such as *finance* would fit in economic anthropology) but may not seem that it would be a core concept within the other categories (such as, for example, in medical anthropology or the anthropology of religion). In this case, place the concept in the nonoverlapping space of that particular literature.

b. *Does the concept fit within the overlap between two or all three literatures?* A concept may be relevant to more than one literature, so place it in one of the mandorlas (overlapping spaces) between two literatures, or within the central overlapping space.

c. *Does a concept seem not to fit well anywhere?* Place it outside the diagram, near the literature it seems to have the most affinity for.

3. *Double-check your work.* When you place a concept within one or more literature spaces but not others, that is a significant and meaningful gesture. It will guide your literature searching, so be sure you've made the best decisions you can.

Now take a moment to listen to your own fully populated Venn diagram. Begin to notice which of your concepts are (1) *unique* to literatures or are *uniquely missing* from the literatures you want to contribute to and which of them (2) represent *overlaps* between two or all of your literatures. The next step shows you how you can begin to connect these two kinds of concepts—*unique and overlapping*—to create a project that makes a productive new contribution to all three literatures.

EXERCISE 4B Create Your Concept Combos

Using a scholarly search engine, you will investigate concept combos that pique your curiosity and seem intuitively promising. Such searches will help you answer these questions: Has anyone else created an ethnographic project that connects these concepts? If so, did they do so in relation to the peoples, places, processes, and contexts that my project features? Is there a justifiable need in scholarly and broader social worlds for a new project like the one I am assembling?

This multidimensioning exercise requires you to perceive your key concepts that show up in the two categorical modes described above (and with which you are already familiar as you've done other concept work). These include those that are unique versus overlapping among your key literatures, as well as those within your Key Concepts table that are more specifically empirical versus those that are more broadly contextual and theoretical. Successful projects maintain a connective tension among these conceptual modes in order to show their significant value. To reiterate, a successful project must show that it is combining concepts in an exciting and productive way. The concepts being combined are not just empirically specific or broadly generalizable but represent a combination of the two.

Here are descriptions of unique and overlapping concepts:

Unique concepts: These include concepts that seem unique to a literature and those that may appear outside the Venn diagram.

Overlapping concepts: Concepts that show up in common overlap spaces between two or all three literatures. These concepts probably already bridge literatures but could be linked with a concept that remains *unique* to a literature or a concept that remains *outside* all the literatures in order to open a space of crosscutting possibility.

Here is a reminder of concepts that can be understood to be more empirically specific versus broadly contextual and theoretical:

Empirical concepts: These concepts in your 4Ps indicate fieldwork-grounded concepts, such as the kinds of spaces you will be in, the groups of people you will be meeting with, the parts/things that people in your worlds are focused on or have a stake in, or the kinds of processes you will be witnessing or participating in.

Broadly contextual/theoretical concepts: These concepts are usually found in your Broad Contexts or Other Concepts sections. These processes or conditions represent far-reaching dimensions of what people experience indirectly or are terms that people in your worlds or scholars use to generalize a category of social experience.

Let's return to McKinson's and Palmer's concept-combo tables (Examples 4.4 and 4.5) and diagrams to listen to what they tell us. The tables present some examples of concept combos that do not follow the combinatory rubric above versus those that do.

Combos with balance provide search terms that allow accurate assessments of the state of the literature in relation to a project's specific fieldwork areas, aims, as well as primary contextual and theoretical scaffolding.

EXAMPLE 4.4 McKinson's Concept Combos

Problematic Concept Combos	Problem	Solution: Combos with Balance
Jamaica + artisans + metalwork	Too empirically specific	Jamaica + artisans + security
Security + Blackness + the State	Too broadly theoretical	Security + Blackness + the State + Caribbean
Materiality + slavery	Too broadly contextual	Materiality + slavery + Jamaica
The State + Black geographies	Too broadly contextual and theoretical	Home security + the State + Black geographies

EXAMPLE 4.5 Palmer's Concept Combos

Problematic Concept Combos	Problem	Solution: Combos with Balance
Peru + Mormonism	Too empirically specific	Peru + Mormonism + migration
Mormonism + migration	Too broadly contextual	
Racism + religious membership + migration	Too broadly theoretical	Citizenship + religion + migration + Peru
Soul + the State	Not empirically specific enough	Soul + citizenship + the State

How can you identify balanced combos? The next steps will help you mobilize your concept-populated Venn diagram to identify three to five possible concept combos to investigate with literature reviews. Follow these instructions:

1. *Listen for crosscutting concept-combo opportunities.* Use your earlier experience with literature reviewing, when you used key terms to search databases. Doing so will help you begin thinking about some possible combos you want to do some searches on. Just as you needed to think about connecting empirical, contextual, and theoretical terms to find works relevant to your project, you will be doing the same thing again. But now, you'll do it in way that is anchored in your project's unique conceptual landscape.

2. *Create three to five concept combos.* Do this by circling three to four concepts for each concept combo grouping. Take time to play around with concepts to see what works. We recommend that you create combos that contain at least one overlapping concept, at least one unique concept, at least one empirical, and one broad context/theoretical concept. Some of these concepts will serve in multiple categories. For example, *citizenship* served as the empirical concept that Palmer was investigating as well as a broader theoretical and contextual concept.

3. *Create your Concept-Combo Literature Search table* (see Table 4.1 for an example). Include a space where you can make notes on what you discover when you do the literature searches. Enter the three to five concept combos you identified. Based on what you find, you may end up expanding this table to include new or improved concept combos, so your eventual table may have more rows.

4. *Use your combos to search for studies that feature these concepts.* If you find some that yield many or very few results, this is not a cause for eliminating this concept combo. We'll prompt you to make some assessments that will help you evaluate which concept combos may be the most promising. In addition, you may begin to suspect that one of your literature categories needs

TABLE 4.1 Concept Combos Literature Search

Concept Combos	Search Notes

adjustment, or maybe there is another literature that your project may also make a significant contribution to. We'll help you assess that as well! Consider taking a twenty-four-hour break before proceeding to the next exercise.

EXERCISE 4C **Do Concept Combo Literature Searches and Assessments**

The goal of this next phase is to assess the multidimensional potential of your concept combos by documenting information about your literature search results. You may have to mix and match combos while you do these searches to land on the combo(s) that best represent(s) your project's contribution.

Before you begin this process, you need to select one or two literature search engines that will provide the most comprehensive sources for this kind of work. During Module 2, you were probably introduced to a variety of others. Take a moment to confirm which search engines are the most relevant to the community of scholars and researchers that you wish to engage. It could be AnthroSource, WorldCat, Web of Science, Google Scholar,

ResearchGate, Academia, or engines available in, and relevant to, your region. Discuss these options with your group members and advisers.

Get ready to use your chosen search engine to search abstracts and titles (we recommend these) or full-text searches for your concept combos. Use a relevant timeframe (such as five to ten years) that will be useful for you, especially when you conduct a more extensive literature review for the purpose of writing a proposal. We do not prescribe what timeframe is relevant to you, because in fact you may need a much longer timeframe—perhaps since anthropology articles began to be archived! We recommend that you discuss this with your advisers and design group. Unlike the literature search process in Module 2, your searches should focus on single ethnographies, not literature reviews.

Taking effective notes about your results include addressing how many results you got with that combo and how much the studies appear to overlap with yours. You can determine this information by documenting the pivotal topical, purpose, and argument-making elements of each piece. Documenting and assessing these citations and your overall sense of what you found is crucial for successful project development and justification. This work will save you lots of time later during the literature review phase of proposal writing.

1. *Search on each concept combo and record your literature findings in your Concept-Combo Literature Citations table (see Table 4.2 for a template).* Draw on the rest of the prompts for further concept combo refinement.

2. *Note the topic, aims, and argument.* Read the abstracts, pay attention to the introductory paragraphs, get clear about the thesis/argument, and understand the conclusions of each piece.

3. *Assess any empirical and/or theoretical overlaps with your project.* If you find significant overlaps in a study or two, you will need to look more closely to see how the piece differs from your work. Pay attention to those studies' conceptual frameworks: Are they using concepts equivalent or different with yours? Do you see the need to fine-tune your concepts in order to distinguish your study? We'll address these issues when you reflect on what you found.

TABLE 4.2 Concept Combo Literature Citations

Concept Combos	Search Results: # of studies and notes	Citations from [relevant timeframe]

4. *Keep track of citations using bibliographic software.* As we indicated, you're going to generate studies that will be very helpful when it comes to completing formal, project-focused literature searches in the near future.

Now you will do some assessments and make some decisions in order to decide which one (or two at most) concept combo is the most promising for your project. We will return to this in Module 7 when you work on designating project significance, so be sure to devote time to this assessment.

5. *Which concept combo situates your project as a unique contribution to the literature?* Follow these prompts to derive your best concept combo. This may entail taking the best terms from two semisuccessful combos and combining them into one powerful combo.

   a. *Assess combos that retrieved no results.* This is tricky. It could mean that no one has studied this conceptual constellation or your search terms were too obscure. Write about your hunch on this and continue to evaluate the next combos.

b.   *Assess combo(s) that retrieved citations.* Among these combos, which yielded citations that pay attention to all of the concepts in the combo? Such combos may not be as promising for distinctive research as those that attend to one or two but not all of the concepts in a combo. Compare these situations and write about which combo(s) appear the most promising. This may mean that the best combo may be a mix and match of the combos that received citations. If so, write about what this new revised combo could be.

c.   *Select one key combo and write about its potential.* Perhaps one of your key concept combos can work as it stands. Or perhaps you have one that simply needs to be terminologically adjusted. Using your responses above, identify and write about which of your key concept combos fits the bill as a multidimensional combo. If you cannot land on one, assess if you need to simply fine-tune one or two of the terms in ways that make it more relevant to how you envision your project and its fieldwork. If you make any changes to your concepts, adjust your Key Concepts table. Or, if adjusting this combo doesn't feel adequately promising or multidimensional, do the next step.

d.   *Assess if you need to adjust your combo or shift to a new combo.* If your combo doesn't seem promising or needs to swap out its concepts for others you have, return to your Venn diagram and draw a circle around another cross-cutting combo that reflects what will make your project distinctive and needed.

In the following prompts you will take time to zoom out to reflect on your key literatures. Sometimes when researchers do literature searches during project development, they intuit that they may not have identified all the actual literatures they will contribute to. Or that they need to substitute one for another. This is a crucial development stage to pay attention to. Multidimensional magic and a new sense of project congruence and significance can come with a literature substitution or addition. You may already have the right literatures, but it's worth taking a moment to check them. See Rahman's reflections on his literature choice for inspiration.

EXAMPLE 4.6 Rahman's Decision about Literature
Substitution and/or Addition

Rahman was faced with a literature substitution or addition decision. Were materiality and digitization simply unique concepts that don't fit neatly within the literature Venn diagram he has, but which can be used to make a crosscutting combo within the three existing literatures? Or does one of his literatures need to be swapped out for another one? Or should he add science and technology studies to his literatures? Rahman's three existing literatures were sound, so swapping one out would have changed the project in a way he didn't want. But he productively and definitively engaged the question of whether to add a new literature or not.

If he chose to add science and technology studies, he would then commit to building a framework of inquiry that would *equally inform* economic anthropology, South Asia, urban anthropology, and science and technology studies. None would be subordinate to the other. This would have created a new shape for the theoretical and ethnographic significance of his project. In other words, with the addition of science and technology studies, Rahman could show that understanding the speculatively digitized "plot" market in Pakistan today cannot be separated from understanding the rise of financial technologies spurred by national and digital securitization. And, by extension, that international real estate market processes today cannot be understood without examining the technical forms that virtual reality, territory, and securitized technologies take in particular city settings.

In the end, Rahman decided that science and technology studies was not a literature he wanted to substitute or add. This decision is based on a core project development issue that everyone has to face. Namely, you will need to decide which literatures you will fully engage in your proposals and thesis, as well as literatures that help define your professional identity.

e. *Assess the congruence between the citations and journal titles with your key literatures.* Do your citations match up with your literatures? In other words, are some of your citations from *Medical Anthropology Quarterly*, which is expected because one of your key literatures is medical anthropology? Sometimes those journal titles will be subtopics related to, but not precisely the same as, your broad literatures. For example, some of your citations on corporate regulation are from business journals, but your key literature is economic anthropology. This may not signal the need for a key literature change. But if your audit here prompts you to question the congruence of your literatures then take the next steps to review your literature categories.

f. *Do you need to substitute or add a literature?* If you suspect a literature may need to be reconsidered, it may make the difference between a project that people recognize as innovative and one that seems old hat! Take a few hours to return to your literature review work from Module 2, Exercise 2b, to conduct a literature review of the possible substitute literature, or even in rare circumstances, to add a fourth literature to your Venn diagram. Upon reflection, your decision could be based on which field you imagine yourself writing articles to contribute to and which one you would like to use to build a professional profile. See Example 4.7 for inspiration about literature substitutions.

EXAMPLE 4.7 McKinson's Concept-Combo Literature Work

McKinson's concept-combo literature work could have yielded citations that were situated in the anthropology of the Caribbean, which could be a broader site category to contribute to rather than the anthropology of Jamaica. Or this could go the other way: her concept combos could yield citations that are mostly situated in the anthropology of Jamaica. Her decision could take into account not only which of these literature categories she wants to contribute to, but also which category might be more relevant to a future job or vocational activity.

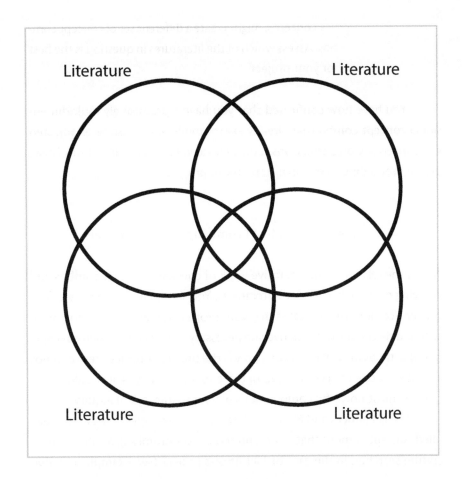

FIGURE 4.4 Venn Diagram with a Fourth Literature

g.   *Update your key literatures and other concept work if necessary.* If you end up with a different set of key literatures, update them. This change in your project elements requires reviewing whether you need to revise or add new key concepts to the Key Concepts table and concept map. If you do, make substitutions or additions, remember to maintain your total number of key concepts at 55. Having made these changes, redo your concept combo work, including populating your Venn diagram. For an example of what a populated Venn diagram looks like with four literatures, see Figure 4.4.

h.   *Reiterate your concept-combo work above.* Explore how your literature substitution, and changes or additions to key concepts, might create a different set of concept combos. Assess which of the literatures in question is the best for your project.

You have now confirmed that you have a promisingly multidimensional concept combo that already exists among your existing or adjusted key concepts. Congratulations! Consider taking a twenty-four-hour break before proceeding to the final steps in this process.

EXERCISE 4D **Identify the Most Promising Concept Combos**

Assess the results of your iterative work of checking concept combos and verifying or changing your key literatures. You should have one to two final concept-combos that revealed that your project is going to be ethnographically and theoretically distinctive and necessary. You may have used concepts multiple times in each combo, so now is the time to contract these combos into at least one concept combo of three to four concepts. It should represent the most powerful combinations of your framework of inquiry.

Using Palmer's and McKinson's searches as examples, review your distilled concept combos that represented the most promising results from literature searches. In the case of McKinson's project (see Example 4.8), two

EXAMPLE 4.8 McKinson's Final Concept Combos

Promising Combos from Searches	Final Combo(s)
Jamaica + artisans + security	Caribbean + slavery + materiality
Security + Blackness + the State + Caribbean	Black geographies + home security + the State
Materiality + slavery + Caribbean	
Home security + the State + Black geographies	

combos showed her contribution to literary conversations about the historical and present material legacies of slavery. It also showed contributions to conversations about the ways that multidimensional Black geographies are structured as city residents and the State mobilize materials and practices differently to produce security. Palmer's project's most promising combo (see Example 4.9) reflects his focus on the multidimensional dynamics of church and state membership that shape migration practices and conflicts.

EXAMPLE 4.9 Palmer's Final Concept Combos

Promising Combos from Searches	Final Combo(s)
Peru + Mormonism + migration	Migration + citizenship + church membership
Citizenship + church membership + migration + Peru	
Soul + citizenship + the State	

1.   *Finalize your combo(s).* Using your most revelatory search combos, cull and combine them to create the final combo(s) to populate Table 4.3.

2.   *Center your combos on your diagram.* Create a new Venn diagram (with your three or four literatures) and place your final most promising concept combo(s) in the middle space (see Figure 4.5 as an example). This is a way to envision the conceptual connections that create the foundation of your framework of inquiry.

This is the concept-work element you will use in Module 5, in tandem with your Key Concepts table and concept map. Follow the prompts in the next exercise to redo or clean these up based on the work you did in this module.

TABLE 4.3 Final Concept Combos

Promising Combos from Searches	Final Combo(s)

EXERCISE 4E **Revise Your Key Concepts Table and Concept Map**

You can now finalize any changes to your Key Concepts table and concept map.

1. *Revise your Key Concepts table.* Now that you've revised your table and taken a look at other studies, it is time to assess all the terms in play. Do some need to be removed? Do some need to be further refined? Remember to keep the total number of concepts in the table at fifty-five. Any more than that tends to make the project unwieldy.

2. *Revise your concept map.* Review and revise your concept map judiciously as needed. Refer to Module 3 to review how to work between your revised Key Concepts table and the map. Following Module 3's instructions, this is again the time to make sure that the concepts are arrayed in a way that represents your project's core, clusters, more peripheral concepts, and all of their connections.

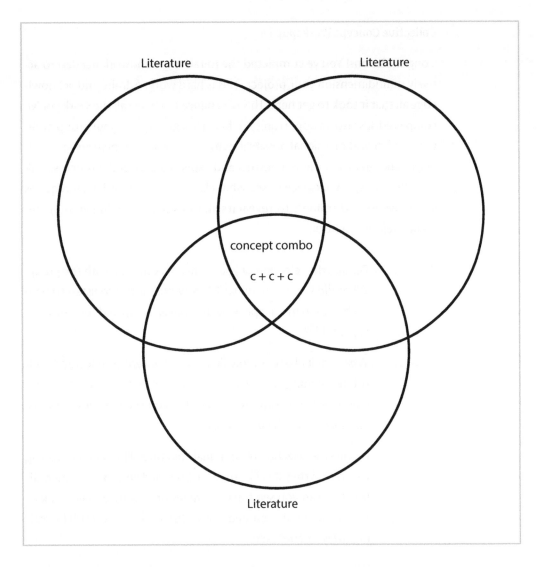

Literature

Literature

concept combo

c + c + c

Literature

FIGURE 4.5 Literatures Venn Diagram with Concept Combos

EXERCISE 4F **Research Project Grid 4**

Revise the literatures portion of your Research Project Grid table if needed. Do a quick written assessment of what the literature work did to adjust your project's overall congruence. Do an audit of your 4Ps and broad contexts as they may have gone through some adjustment while working through this module. Update the Key Literatures, 4Ps, and Broad Contexts, found in the upper half of the grid.

Collective Concept Workspace 4

Congratulations! You've completed the iterative groundwork needed to actively multidimension your project. This is hard work! Admire and acknowledge all that it took to get here. This is a chance to review all the work you've completed. It's also an opportunity to begin imagining with your group other potential multidimensional possibilities for your project. If there are still challenges and fuzziness, this is a normal and expected part of the process. Sink into the concept workspace to see what clarity may arise and what requires spaciousness and stillness to reveal itself. Consider reviewing your agreements before you begin.

1.   Discuss the selection of your concept combo(s) with the group. What did you come up with? How does it feel to listen to how each combo might speak to the concept map and the Key Concepts table?

2.   What was it like to explore literatures, concept combos, and multidimensioning, using the Venn diagram as a workspace? Solicit feedback from your group members on the concept combos' relationship to the literatures.

3.   Discuss the selection of your final literatures. How are they feeling to you? Do they feel like they are bounded and compartmentalized? Or can you see the beginnings of how they may speak to each other in unbounded ways? Discuss what may still be feeling fuzzy or confused.

4.   What were the acts of iteration or multidimensioning that most prompted you to rethink the project's major connections? What lingering concerns do you have?

5.   The concept work you have just completed will go through more changes as you learn more about your project over time. Talk about how it feels to be flexible with future project changes and adjustments. Flexibility is key to flowing with the multidimensional concept-work process.

MODULE 5

Describe Your Research

Now that you have identified your project's initial multidimensionality, you are ready to write a research description (RD). The RD is a 100- to 150-word narrative rendition of what your project is about: what you will do, where, when, with whom, why, and with respect to what kinds of disciplinary objectives and ideas. As a distillation of your project's aims and activities and a showcase for your innovative research concept combo(s), the RD represents your project's core conceptualization and multidimensionality. A strongly crafted RD will help you to identify a fully connective multidimensional object (MO) (Module 6), to design effective research questions, and to articulate the significance of your project (Module 7).

The pithy and cogent RD you develop here will be very useful even beyond the design process. A well-crafted RD can serve to anchor all your project's short and long narrative forms. It can support brief descriptive needs, including project summaries, abstracts, keyword lists, and elevator pitches. When it comes to writing longer documents, such as full proposals, grants, and budget justifications, a strong RD is a touchstone for elaborating a more comprehensive description of your research plans and aims. Being able to create and tailor a succinct research description is an essential professional skill.

Develop a succinct research description that includes broad key concepts; clarify your project's core multidimensional form and possibilities.

In our experience, the hardest thing about writing an RD is to make sure that it clearly narrates the project's main objectives as well as its multidimensional uniqueness. Moreover, it should communicate the project's overarching ethnographic and theoretical structure by giving a sense of its most innovative conceptual relationships. Researchers starting a new project often find it difficult to discern which concepts represent key project features and connections and which are detail terms that can be included in the body of a longer narrative. We'll help you begin this process of discernment by looking at other scholars' succinct project descriptions. Then we offer a thought-provoking exercise that clearly illuminates the heart and soul of your project by reconnecting you with people and communities that inspire your work. Following these introductory exercises, you'll draft your first RD and refine it by comparing it with some example RDs from other projects in process. Lastly, you will learn how to narrate the various facets of your project's multidimensionality.

A successful multidimensional RD's empirical and theoretical concepts are balanced and coherently related. You are familiar with balancing these kinds of concepts in a research imaginary and on a concept map. Here we help you do this in a comprehensively contracted narrative. Finding conceptual balance ensures that the diverse elements that you are working with—such as the people and processes you want to research, the literature concepts you found, and the overarching contexts of the project—"read" as understandably multidimensional. In other words, anyone who reads your RD will see how and why you are putting things together in new or unusual ways but will also be clear about what you're going to do, where, with whom, and why.

The process of honing your RD will help you cultivate project listening and facilitate your capacity to discuss the project with communities important to you, including colleagues and those whom you aim to serve with your work. At the end of this module, you will have a description that you can successfully adjust, as needed, because you will understand what it can and must do to support intellectually and socially responsive multidimensional work.

One last introductory note: we know that it will feel difficult to contract your previous expansive conceptual and graphical work into a short narrative RD, but doing so is highly rewarding. It will help you focus your intentions and efforts. This is the moment when you really begin to see whether your overall project is too broad or too specific, if it includes the most ethnographically and theoretically effective collection of key concepts, and if you are able to communicate its potential multidimensional significance. But don't worry! You will indeed produce a sturdy RD. Its processural refinement will produce a sense of clarity and accomplishment that keeps you securely connected to the potentials of your work.

Exercises

EXERCISE 5A Project Listening: Admiring Successful Ethnographic Descriptions

We want you to take two days off before you begin the RD writing process. During this time you can do this intermediary exploratory exercise, which prompts you to reflect on work that inspires you. This exercise asks you to consider what a project is really about. How do people sum that up in a nutshell?

During your two days off, select three fully fleshed ethnographic works that you admire and that stay on your mind. Think of these as the books or visual works (such as films) that you would want access to if they were the only ethnographic works you could ever access again. They may or may not have something to do with your project, topically or conceptually. But they will work for this exercise if they are written or are visual forms that say "ethnography" to you and that you feel an intellectual and intuitive affinity for. Put them on a real or virtual shelf.

Check out their most distilled project descriptions, which come in various forms: usually a title, abstract, cover or publicity image, description on the back of a book, or film synopsis. Begin to consider what these distilled descriptions tell you—and don't tell you—about the projects' main features and aims.

You will now free-write answers to these questions. There is a prompt at the end of the exercise that may inspire you to revisit and perhaps even revise your research imaginary, concept map, or Key Concepts table, so be sure not to skip it!

1. *Reflect on the descriptions of these works.* Find the publishing or website description or abstract of each work. The aim here is to find a short 100- to 200-word description of what the work is about. That description may not be complete in that it may not be clear about the main argument or its relation to other ethnographic work. Nonetheless it should give you a sense of the project's uniquely arrayed and combined project concepts (as we define them) and its intellectual and social aims. If the description contains an argument statement—a statement about what the project concluded—ignore it for now. When you are done with fieldwork and ready to produce products from your project, you'll be able to write an RD that contains an argument statement. For a project that has yet to be conducted, your RD simply needs to be a concise description of the key concepts and general aims.

   Take a look at each description you have found, and reflect on these prompts:

   a. How does the description's structure flow, in terms of representing the who, what, when, where, and why of the work?

   b. Does the description adequately convey the aims and significance of the work as you know it, having read or seen it? Why or why not?

   c. Does the description reveal how the project relates its concepts in innovative ways? If so, how?

2. *Reflect on visuals.* Look at any cover or promotional visuals for the project.

   a. What do these images convey about what the work is about? Do you think the images are effective or do they miss an important project dimension?

b.   What image(s) might you select to convey what your project is about? Why would you select this/these images? Does each image contain elements that aren't currently in your research imaginary, concept map, or Key Concepts table? Consider revising these if you hit on something important!

3.   *Connect your reflections.* Reflect on the overall sense you have of each project. Do the descriptions and visuals convey adequate information on what the project is about? Why or why not?

4.   *Reflect on the purpose of project descriptions.* What have you found out about descriptions? What makes them effective and why?

Keep these favorite works on hand during the design phase and after. Imagine a work of yours sitting next to them someday! This can help you reconnect with your desire to do a project and to situate it in relation to other works within broader creative spaces.

EXERCISE 5B **Warm Up to Communicating Your Project**

We mean "warm up" in two senses. This exercise helps you lean into what it means to communicate your project clearly. It is also about getting in touch with your warm, heartfelt aims to communicate the core features and value of your work.

Unlike other exercises, do this one *without looking back* at the concept work you've done so far. This exercise is "off the cuff" and mobilizes your intuitive and multidimensional sensibilities; it also cultivates attention to your audience and readers.

1.   *Imagine a meaningful audience.* Find some quiet time to imagine three different people who matter to you; they may still be with us or they may have passed. We suggest the following:

     –   a family member or community elder (like a grandparent or personal mentor figure)

     –   an inspirational scholar (living or passed away)

     –   a spiritual, political, or ethical leader you deeply admire

To begin, find some stillness and let yourself feel connected with them; here are some prompts:

a. *Evoke your persons one by one.* Bring each person to mind as clearly as you can.

b. *Relate to them.* Feel into your relationship with them (whether you know/knew them personally or not). Remember how and why they matter to you. Consider their importance to others as well. This may evoke a variety of emotions. Please give yourself time to sit with and honor the feelings, such as joy or sadness, that arrive.

c. *Feel their experiences and questions about the world.* When you are ready, think about what is or was meaningful to them and, importantly, what they are or were curious about. Think of how they prefer(ed) to know and engage the world.

d. *Reflect and connect.* Write about how each person inspires you and motivates your scholarly inquiries and creative work.

2. *Write three tailored project descriptions.* Write, for each person you chose, a twenty-five-word explanation of what your project is about. As you can already imagine, you will probably use somewhat different terms to render these personally tailored versions. Here are some steps we advise for this process:

a. *Invoke the experience of describing your project.* Start with one of the people and imagine telling them about your project. Be aware of the terms and language you are thinking in.

b. *Write.* Just jot out what comes to mind, without overthinking or hesitating, in the terms or language that feel right.

c. *Edit.* Revise a bit to keep your description to twenty-five words, but it's fine if you go a bit under or over.

d. *Repeat* the process with the next person.

These project descriptions are from the heart. They are connected to individuals who matter to you, but they also may reveal something new to you about the ethnographic who/what/where/when/why of the project that you haven't necessarily felt yet. What you write may only address your project, or they may connect with other deeper and broader visions you have of why your work matters to you. To help you see the freedom and spaciousness available in this activity, Example 5.1 contains our twenty-five-word descriptions for our first book projects.

EXAMPLE 5.1 Olson's and Peterson's Twenty-Five-Word
Project Descriptions

Valerie Olson—*Into the Extreme*

Family: Godmother

I'm studying how US environmental politics and inequalities on Earth are being shaped by what governments and companies do in outer spaces.

Anthropologist: Marilyn Strathern

I want to show how the Western "system" concept has become a political technology to connect more-than-Earthly spaces and things to US social and colonial worlds, and how it extends political ecologies below and above Earth's surface.

Ethical/Spiritual Leader: Thich Nhat Hahn

I want to study how different institutions in the US work to move bodies off Earth; I want to understand the many ways people strive for ways to control the contradiction of being connected and separate from the space of experience.

Kris Peterson—*Speculative Markets*

Family: Mother

I'm studying how the global pharmaceutical industry works in, and exploits, the Nigerian and West African drug market.

Anthropologist: Diane Lewis

I'm studying how racialized, post/colonial forms of finance capital and national banking and economic reforms simultaneously constructed a

---

relationship between fake drug chemistry and market collapse in the making of Nigeria's contemporary pharmaceutical market.

Ethical/Spiritual Leader: Prentis Hemphill

I'm interested in how Nigerians cope with challenging health outcomes and low-quality pharmaceuticals that are direct racial outcomes of World Bank-imposed market collapse.

After you write the descriptions, take a moment to reflect on them, seeing the differences and similarities between them.

3. *Write a reflection about what you learned using these prompts.*

   a. *Note similarities.* What do the descriptions have in common? Consider language, tone (personal vs. formal), terms (generally understood vs. disciplinary), and scope (personal vs. contextual, specific vs. generalized, embodied vs. social).

   b. *Note contrasts.* How do they differ? Again, consider language, tone (personal vs. formal), terms (generally understood vs. disciplinary), and scope (personal vs. contextual, specific vs. generalized, embodied vs. social).

   c. *Reflect.* What do these versions help you learn about your project's core liveliness? How do you express concepts or aims in this exercise compared to how they are expressed in your research imaginary or concept map?

4. *Iterate your concept work.* Now go back and retrieve your most updated concept map, Key Concepts table, and multidimensional concept combo(s). See whether you want to alter any terms for your concepts. For example, you might decide to express them in more socially specific or straightforward ways that might strengthen the communicability or relevance of your project to scholars as well as people and communities who are important to you.

This exercise is good to come back to if you get stuck writing your RD, or you find yourself spinning out into chains of abstractions or jargon. You can always "ground-truth" your RD by coming back to these simple but multifaceted descriptions.

EXERCISE 5C Understanding Research Description (RD) Structure

We begin by showing you examples of Key Concepts tables, concept combos, and RDs, based on the work of two researchers we worked with (see Example 5.2). You will then reverse engineer example RDs to get a feel for structure, flow, and organization.

Notice how each RD contains *all* the concept-combo terms, as well as a select collection of concepts from the concept map. We call these concepts

EXAMPLE 5.2 Haven's and Wilkinson's Key Descriptors, Concept Combos, and RDs

Forest Haven's Project	Research Description
Concept combos: Settler colonialism + sensing Indigenous + food + Alaska + sensing **Key Descriptors:** *Research problem descriptors*: Indigenous food practices; Alaska; settler colonialism; bodily sensing; technological sensing *Empirical descriptors*: Alaskan Native Peoples; state authorities; Indigenous food; sensing *Contextual descriptors*: sensory regulation, settler colonialism, assimilation, State-Indigenous conflicts *Theoretical descriptors*: sensing; sovereignty; indigeneity, colonialism *Research aims descriptors*: food sensing, everyday life, technical practices, settler colonialism, sovereignty	Given ongoing conflicts between state authorities and Alaskan Native peoples, this project examines how twentieth-century assimilation policies still inform the state regulation of Indigenous food practices. It focuses specifically on the sensory dimensions of Native food gathering, hunting, and processing. It positions sensing as a site through which forms of knowing and sovereignty are negotiated and enacted. In this way, the project can examine food sensing at the levels of everyday embodied Native food practices and the extreme technical practices of Alaskan state authorities regulating Native plants, animals, peoples, and practices. This project aims to analyze the persistent strategies of Native people to respond to shifts in the ongoing project of settler colonialism.

Annie Wilkinson's Project

Concept combo:

Mexico + family + security + right-wing

Key Descriptors:

Research problem descriptors: Mexico, right-wing movement, gender discourses, security

Empirical descriptors: right wing activists, Mexico, security policies, "gender ideology," conspiracy theories

Contextual descriptors: global populism, violence and democracy, securitization, profamily legislation, corruption

Theoretical descriptors: gender, security, equality, sovereignty

Research aims descriptors: gender conflicts, security, regional populist movements

The movement against "gender ideology"—a concept used by right-wing groups globally to reference the social construction of gender—is gaining force across Europe and Latin America. Over the past decade, profamily movements have claimed that gender is rooted in nature and that this scientific fact must be defended. This allows right-wing organizations in Mexico to claim that they must secure the family against gender equality and LGBTQ rights. In doing so, the Mexican pro-family movement makes "gender ideology" a national security problem that bears on state corruption, violence and crime, and neocolonialism. Using ethnographic research, the project follows how right-wing theories of the family connect to theories of the state. It also investigates how gender and security agendas are being shared among authoritarian populist movements across the globe.

the *key descriptors*. This multidimensionally robust and judicious selection of concepts yields a balanced and inspiring representation of the ethnographic and theoretical structure of the project.

These two RDs demonstrate an effective sentence-by-sentence rollout of each project's multidimensionality, key descriptors, main ethnographic problem and features, broader contexts, and research aims, as well as an indication—even if very cursory—of overall project significance. Examples 5.3 and 5.4 demonstrate the main components and sentence order that structure these example RDs; however, you are welcome to experiment with your own structure. The only requirement is that your RD should feature most, or ideally all, of your concept-combo terms.

RD Structure = Research Problem + The Problem's Key Empirical, Contextual, and Theoretical Descriptors + Research Aims

---

EXAMPLE 5.3 Haven's RD Structure

Research problem you are investigating in terms of the main site(s), subjects, and processes.	Given ongoing conflicts between state authorities and Alaskan Native peoples, this project examines how twentieth-century assimilation policies still inform the state regulation of Indigenous food practices.
The Problem's Key Empirical, Contextual, and Theoretical Descriptors: the specific social and cultural features and relations, space/environment, power, etc.	It focuses specifically on the sensory dimensions of Native food gathering, hunting, and processing. It positions sensing as a site through which forms of knowing and sovereignty are negotiated and enacted. In this way, the project can examine food sensing at the level of everyday embodied Native food practices and the extreme technical practices of Alaskan state authorities regulating Native plants, animals, peoples, and practices.
Research aims in terms of the disciplinary/intellectual stakes and emerging social stakes.	This project aims to analyze the persistent strategies of Native people to respond to shifts in the ongoing project of settler colonialism.

EXAMPLE 5.4 Wilkinson's RD Structure

Research problem you are investigating in terms of the main site(s), subjects, and processes.	The movement against "gender ideology"— a concept used by right-wing groups globally to reference the social construction of gender—is gaining force across Europe and Latin America.
The Problem's Key Empirical, Contextual, and Theoretical Descriptors: the specific social and cultural features and relations, space/environment, power, etc.	Over the past decade, pro-family movements have claimed that gender is rooted in nature and that this scientific fact must be defended. This allows right-wing organizations in Mexico to claim that they

	must secure the family against gender equality and LGBTQ rights. In doing so, the Mexican pro-family movement makes "gender ideology" a national security problem that bears on state corruption, violence and crime, and neocolonialism
Research aims in terms of the disciplinary/intellectual stakes and emerging social stakes.	Using ethnographic research, the project follows how right-wing theories of the family connect to theories of the state. It also investigates how gender and security agendas are being shared among authoritarian populist movements across the globe.

To help you understand how effectively structured RDs strategically build on key descriptors, we invite you to reverse-engineer two RDs, which we derived from other researchers' work (see Table 5.1). By "reverse-engineer" we mean taking the RD apart to see the specific key concepts being used in the description.

1. *Reverse Engineer the* RD. Using your best guesses, place single words or phrases next to the corresponding key descriptor type. There's no right answer to this, we're simply giving you an opportunity to think about how an RD is constructed out of conceptual parts. You can work alone on this, or you can repeat this in your group using some of the group members' favorite

TABLE 5.1 Reverse-Engineering Kladky's and Badami's Research Descriptions

| Ellen Kladky's Project Key Descriptors
Research problem descriptors:
Empirical descriptors:
Contextual descriptors:
Theoretical descriptors:
Research aims descriptors | RD
Within deindustrializing Appalachia, there is a growing movement to manage family life using religious financial practices. This project examines intertwined changes to |

	financial subjectivity, family life, and social class through an ethnography of Christian financial literacy programs in the region. Based in working class communities in economic transition, it follows what happens to white identity and family processes as communities lose historical forms of livelihood, take on debt, and reorient to racial and class privilege. Situated in broader interconnected opioid, environmental, foster care, and job safety crises, the project examines how novel economic practices are changing the concept and experience of class in the United States.
Nandita Badami's Project **Key Descriptors** *Research problem descriptors:* *Empirical descriptors:* *Contextual descriptors:* *Theoretical descriptors:* *Research aims descriptors:*	**RD** India has become the second largest democratic solar market in the world. The appearance of sunlight as an industrial energy form in India intersects with its historical presence as a religious power source. As a result, the politics of sunlight in the contemporary context of Indian nationalism is both physical and metaphysical. Situated in the outskirts and interior sections of Mumbai among policymakers, solar technicians, solar smugglers, off-grid builders, and spiritual centers associated with the sun, this project investigates how India's contemporary political conflicts are shaped by people's different engagements with sunlight's promising "alternative" characteristics as clean, free, and unlimited. By investigating Indian solar power as a contested moral and political force in villages and cities, the project illuminates the combined cultural and economic dimensions of national alternative energy transitions.

ethnographies that you each worked with in the first exercise of this module.

2.   *Reflect on your responses.* Notice how an RD contains only the most *relevant* descriptors from any Key Concepts table. How does using those descriptors in narrative form powerfully convey your project's structure and intentions? You can free-write about this or, again, discuss in groups. In what follows, we help you identify these key descriptors.

EXERCISE 5D **Draft a Baseline RD and Identify Key Descriptors**

You will first draft a baseline RD to work with. Then you'll refine this description by populating it with carefully chosen key descriptors. After this, we'll help you play with how you roll out and relate these concepts, how to see multidimensionality, and how to make your RD sentences flow.

1.   *Draft a baseline RD using the RD structure guide.* Don't look back at any of your previous concept work yet. Using the sentence order template in Example 5.3 or 5.4, draft a quick RD, just off the cuff.

2.   *Edit as necessary to keep it pithy.* Keep the description under 150 words.

Now it's time to see if you've got most if not all of the key descriptors you need in this draft RD. The best way to do this is to take a moment to judiciously contract your Key Concepts table into a set of key descriptors. You did this kind of tailored contraction in Module 2 when you reduced your Key Concepts table to keywords for the purpose of facilitating literature reviews, as well as in Module 4 when you built on that work to create your concept combos. In addition to the concepts in your combos, you need to focus on this important question: What are the concepts that represent the topical core and major aims of your project? In other words, what are its *descriptively vital concepts*?

The concepts that you select must be drawn from every major category you've identified in your table. That is, they must illuminate the project's specific empirical features as well as its most important contextual and theoretical concepts. We'll help you check that balance in a bit.

While identifying your key descriptive concepts, you may have to revise some of your conceptual terms and phrases, and that's fine. If you do make some tweaks as you work, remember to update your Key Concepts table accordingly.

3. *Gather concept-work elements and prepare to identify your key descriptors.* Put your Key Concepts table next to your concept map and your literatures Venn diagram. See if you can identify some vital descriptors simply by looking at how you've placed or listed them in these concept-work elements. For example, look at the concepts that you've listed first in each of the concept-category boxes in your Key Concepts table. On the concept map, look at the concepts you may have represented as "standout" or "core" concepts in some way. You can bold them or make them larger than other concepts by clustering them in the center, or by any other means of visual distinction. On the Venn diagram, look at your concept combos and remember their significance as key terms that indicated how your project was going to be innovative. Consider whether the positionality or emphasis of the concepts in your table, map, or Venn diagram is giving you any clues about whether these are the main project elements, versus other concepts that could function as supporting concepts. Positionality or emphasis may not provide those clues, but it can help you with the next steps.

4. *Identify your key descriptors.* We know that all your key concepts are important, but you simply can't use them all in a short description! Your aim is finding the most vital concepts to use. Get five highlighter pens ready because you will be identifying three to five vital concepts that fit into five categories. You will then narrow them down into a manageable number with which to write your RD.

    a. *Problem or aim descriptors:*

        i. Looking at your table, mark (circle, highlight in one color) three to five concepts that point to the most significant social change, problem, paradox, or emerging or persisting dynamic that you are investigating. For example, the project may center on a crucial historical shift or a recent or upcoming event. Or it might involve the presence or absence of a form of life, such as a new identity, a category of life, a changing way of relating, or emerging human/more-than-human dynamics. After you circle those concepts, look at the remaining concepts. Do you find yourself hesitating about leaving out any unmarked concepts? If so, go back to the concepts you marked as key problem descriptors and see if they need to be exchanged with another concept. You may need to do this several times until your two to three problem concepts seem right.

ii. See if you can use those concepts to write a phrase to describe your research problem. For example, "By 2030, more than 70% of North American groundwater will be too salty to drink without filtration, opening a future of salinity filtration contamination and inequality."

b. *Use another highlighter color to mark the most vital three to five empirical descriptors.* By *empirical* we mean concepts related to your 4Ps, although they can also be found in your other concepts. Ask: What are the people, parts, places, or processes are most central to your project's problem or aim? Identify your descriptors by circling or highlighting them. After you do this, conduct an audit of concepts you choose versus those you didn't to ensure that you chose well.

c. *Use another highlighter color to mark the most vital three to five contextual descriptors.* By *contextual* we mean concepts that refer to broad defining circumstances or processes in which your 4Ps are located. Ask which concepts clarify the scope and importance of this problem in terms of interrelated or conflicting broader contexts. Identify your descriptors by circling or highlighting them. After you do this, conduct an audit of concepts you choose versus those you didn't to ensure that you chose well.

d. *Use another color to mark the most vital three to five theoretical descriptors.* By *theoretical* we mean generalizable concepts. To find these, ask which concepts (often from the literature review or other intellectual or social discursive domains) best describe what you aim to theorize or how your project brings ideas or theories or literatures together. Or ask which descriptors indicate how the project is contributing to or challenging the intellectual and conceptual landscape. Identify your descriptors by circling or highlighting them. After you do this, conduct an audit of concepts you choose versus those you didn't to ensure that you chose well.

e. *Find three to five research aims descriptors.* By this we mean concepts that indicate how your project will achieve your aim of illuminating the social and political implications of what is happening to whom, where, and what. In other words, why is this project relevant to understanding what is happening in social worlds? Identify your descriptors by circling or highlighting them. After you do this, conduct an audit of concepts you choose versus those you didn't to ensure that you chose well.

f. *Ensure you RD gestures to or invokes your literatures.* We want you to place your concept combos in your table because your RD needs to demonstrate that you have searched the literatures to establish the validity of your conceptual framework. RDs don't have to state what your literatures are specifically, but they do have to invoke the broader scholarly parameters of your project. Your concept- combo terms, which resulted from your careful searching, do this invocation work for you.

To compare your descriptor identification process with examples from projects we've already examined, take time to review Examples 5.5 and 5.6.

EXAMPLE 5.5 Haven's Key Descriptors

Research Problem Descriptors	Native subsistence food practices; state regulation of food practices; sensory regulation and control; settler colonialism
Empirical Descriptors	Alaskan Native peoples; state regulators; Indigenous food; senses; technologies and laws
Contextual Descriptors	Sensory regulation and control; historic and contemporary conflicts over food and space; environmental laws, sovereignty
Theoretical Descriptors	Food materiality, sensing, colonialism, embodiment, the State, indigeneity, sovereignty
Research Aims Descriptors	Understand the contemporary relationship between sensory politics, food spaces and access, and the ongoing assimilation project of settler colonialism

EXAMPLE 5.6 Wilkinson's Key Descriptors

Research Problem Descriptors	Mexican pro-family organizations, transnational pro-family organizations, pro-family activism, state securitization activities
Empirical Descriptors	Right-wing activists and organizations; antigender ideology discourses; pro-family organizing; right-wing theorization, policies

Contextual Descriptors	Transnational right-wing movements; popular security concerns; Mexican security state; social equality; centering the family
Theoretical Descriptors	Gender; security; the right; conspiracy
Research Aims Descriptors	Understand links between opposition to the concept of gender, limits on sovereignty and equality, and right-wing populist movements

Now that you have identified your key descriptors, load them and your concept combos into a Key Descriptors table (see Table 5.2).

You've just done a lot of project contraction work. Before you move on, consider taking a twenty-four-hour break to come refreshed to the next exercise, which will take some conceptual effort.

EXERCISE 5E **Interrelate Key Descriptors**

This exercise narratively expands and expounds the multidimensioning concept-combo work you did in Module 4 by going beyond the $x + y + z$ combo form. It helps you convert your project's conceptual multidimensionality into sentences.

To get there, we will start by creating key descriptor combos that are different from the literature review concept combos. Those *literature review* concept combos created bridges to focus your attention *externally*. They helped you to check how your project's key conceptual dimensions relate to *other* projects' key conceptual dimensions. The *key descriptor* combos focus you *internally*. They help you compose sentences that articulate your project's multiple dimensions. You will use these multidimensional articulations to refine your draft RD.

Research description finalization is a pivotal moment in multidimensional research design. Articulating multidimensionality in your RD allows you to begin to express in sentences the innovative, urgent, and intellectual

TABLE 5.2 Key Descriptors

Problem or Aim Descriptors	
Empirical Descriptors	
Contextual Descriptors	
Theoretical Descriptors	
Research Aims Descriptors	

or socially significant contours of your project. At this point you may have only one or two multidimensional facets, but that's actually all you need.

Before you begin, recall our discussion about multidimensionality of projects by Tariq Rahman, Jason Palmer, and Kimberley D. McKinson. Take a moment to remind yourself about what multidimensioning is and how it works to make creative conceptual relations perceptible.

Multidimensioning: The process of defining a project's conceptual combinations and using them to create congruently integrated project elements, from a research description to research questions.

Let's return to Forest Haven's and Annie Wilkinson's projects (in Examples 5.7 and 5.8) to narrate some of their multiple dimensions as you prepare to articulate your own. Look at the key descriptor combos and how they can illuminate the different combinatory and conflictual dynamics that these projects are calling attention to.

EXAMPLE 5.7 Haven's Key Descriptor Combos and Multidimensionality

Key Descriptor Combos	Multidimensionality
Alaskan Native peoples + Indigenous food + senses Indigenous food sensing + settler colonialism + technical sensing sensing + sovereignty + food	*Food sensing multidimensionality:* Food's sensory materiality doesn't begin or end with its features as a nutritional substance or agricultural commodity. Food sensing is a process that is meaningful at various material and spatial dimensions. This is evidenced by differences in how Alaskan Natives and state officials enact sensing-based forms of identity and relationality. *Food politics multidimensionality:* There are two regimes of food "sensing" being put into political conflict by the state: Indigenous "subjective" forms of sensing food vs. the State's "objective" forms of sensing food as environmental resources and commodities. *Food sovereignty multidimensionality*: Connecting sensing and sovereignty helps to illuminate the meaning of food governance and dispossession on Native terms. Studying how sensory politics are not just "local" or "personal" bodily politics, but connect to larger spatial and technical sensory processes that shape processes of sovereignty and ongoing colonization.

EXAMPLE 5.8 Wilkinson's Key Descriptor Combos and Multidimensionality

Key Descriptor Combos	Multidimensionality
Mexican security state + limits on sovereignty and equality opposition to the concept of gender + the family + state security concerns + transnational right-wing social movements	*Security multidimensionality*: Mexican pro-family movements naturalize the link between personal and national security by making social equalities a threat to both. Studying how this works adds dimensionality to understandings of national security by following it as a bidirectional process extending between family and nation-state. *Gender multidimensionality*: Gender is examined as a transnational securitized concept (by studying the production of naturalized sex and sexuality discourses and antigender discourses). In this mode, it allows inquiry into how gender discourses can become interchangeable with senses of security that resonate within and outside of Mexico.

Playing with descriptive combos and narrating the kinds of multidimensioning that is happening in a project takes time. Working with how to narrate your multidimensioning at this early stage will help you work your way toward longer forms of project description and justification. Here are some prompts to find multidimensionality within and across some of your concept categories:

1. *Select concepts that are multidimensionally unique.* Use your Key Descriptors table to check *within* and *between* concept categories. In our examples, Forest Haven brings ways of sensing food and colonial regulatory governance together, and Annie Wilkinson brings the gendered family and national security together (see Example 5.9).

EXAMPLE 5.9 Haven's Multidimensioning within a Category

With respect to her project's 4Ps, Haven connects "Alaskan Native peoples + Indigenous food + senses" to explore how Indigenous modes of sensing food in Alaska, as in other places, has multiple dimensions because food isn't a singular material and sensing isn't a uniform mode of experience.

a. *Within a category:* Explore how you can connect and narrate your project's promising or novel interrelational features within the empirical, contextual, or theoretical (see Example 5.9). What intracategorical conceptual aspects are you bringing together in an innovative way?

b. *Between categories:* Look across table categories. Where else do you see multidimensionality? Or where can you see that you could make it happen in terms of your overall aims or the social or political meaningfulness of your project?

Now that you have created key descriptor combos, fill out table Table 5.3. Build on some of the concept-combo work and reflections you

TABLE 5.3 Key Descriptor Combos and Multidimensionality

Key Descriptor Combos	Multidimensionality

generated in Module 4, taking the opportunity to refine these articulations and put them more succinctly.

At this point, it's important to take a break before you move on to edit your RD.

Before you refine your draft RD and, later finalize it, think back to our discussion about how book jackets distill books into their most vital elements
and interesting dimensions. Your goal is to create a stand-alone description,
a good-enough summary. When you read it, you know basically what the
project is about.

Here are some dos:

- *Do address a general audience.* Whatever you do at this point,
  don't make it sound like a well-polished or elite abstract! At
  this stage it can have an exploratory feel; bring some casual
  language to it.

- *Do keep it loose and engaging.* Again, this exercise is not about
  crafting an airtight RD. It is about anchoring the project's main
  reason for being, even if a full sense of that is still not apparent
  to you. The continuing, recursive work on the RD, eventually in
  relation to the multidimensional object, will help gel the project over time.

Here are some don'ts:

- *Don't list specific literatures.* You can write that into a literature
  review. This is about making an overall project frame.

- *Don't include methods or analysis plans.* You don't need to detail
  how you are going to do the research (methods or analysis). This
  is an ethnographic project, so it is a given that you will do fieldwork, interviews, observations, and other innovative methods.

- *Don't pose questions.* Don't pose any data-gathering or interview
  questions. As we keep mentioning, this is an exercise in beginning to express the scope and stakes of the project. This work
  will prepare you for answering the big "so what?" question in
  Module 7, which we call the *scoping question.* For now you're
  building the frame of the house, not the roof.

Relax into refining your draft RD. You will have opportunities to reiterate
your RD in other modules, so hold it with an open hand. Then, keeping the

previous dos and don'ts in mind, assess and revise your draft RD following these prompts:

1. *Compose the RD using key descriptors.* Edit your original draft RD, making sure you include as many key descriptors as possible, keeping in mind the multidimensional aspects you may have surfaced in the previous exercise.

2. *Assess the balance of key descriptors.* Decide on a color code for your RD's descriptors (Problem or Aims, Empirical, Contextual, Theoretical, Research Aims Descriptors) and mark your RD accordingly. Perceive how the RD looks: Are you contextual or theoretical concept heavy, with a lack of empirical specifics (people, parts, processes, places)? Most people lean toward empirical or contextual/theoretical. Here are examples of sentences about the same topic that lean either way:

   Empirically specific: "This study examines how and why North Dakota farmers are not growing wheat or vegetables under the current US political administration."

   Contextually and conceptually broad: "This study will illuminate changes in transnational commodities trading and its relationship to agricultural growing patterns and election cycles."

A good RD has a balance between both kinds of descriptive narration, so now is the time to perceive what kinds of descriptors are missing or underrepresented and correct that accordingly.

3. *Audit for empty or hypergeneralized placeholder terms.* Now that you can see your broad contextual and theoretical terminology, make sure your RD isn't overloaded with concepts that can apply across many, if not most, projects (like *neoliberalism, development, globalization, transnationalism, the body,* etc.). If you're using many such abstract theoretical terms to convey foundational meaning in the RD, consider and feel into where empirical specification may be necessary.

4. *Put your sentences in order.* It's time to think about conceptual structure. You can highlight the project's multidimensionality at

any juncture, showing the dynamic connections between empirical, theoretical, contextual, or other concepts that are new or different. But often it's not immediately clear how to roll out the project's components, one by one. Drawing on the RD formula can be helpful to check for narrative flow: RD Structure = Research Problem + The Problem's Key Empirical, Contextual, and Theoretical Descriptors + Research Aims.

–   **Research problem** you are investigating in terms of the main site(s), subjects, and processes.

–   **The Problem's Key Empirical, Contextual, and Theoretical Descriptors**: the specific social and cultural features and relations, space/environment, power.

–   **Research aims** in terms of the disciplinary/intellectual stakes and emerging social stakes.

To help you see what this looks like, return to Examples 5.2–5.8 to assess how others reproduced or modified this flow.

5.   *Keep it pithy*. Stay within the word limit. If you need help whittling down or padding your RD, this might be time to get help from peers in group work.

Congratulations! After doing these exercises, you have a strongly drafted RD! Now that you've tried, you know that it's hard to do! We acknowledge this. You will need to tweak the RD as you move through the concept work and we will indicate opportunities for this along the way. Don't worry if you got stuck or struggled or even couldn't do it.

EXERCISE 5G Research Project Grid 5

Insert your RD into the Research Project Grid table. This is a big moment because you can begin to see clearly the succinct baseline foundation of a strongly congruent project. You can see the directly perceivable relationship in the table between 4Ps, Broad Contexts (the who, what, where, and how), Key Literatures (the world of project analysis and communication as well as disciplinary theorization) and RD (your empirical and analytic aims). This is the relationship to come back to over and over, all along the way.

Collective Concept Workspace 5

At this stage, you have worked with lots of (and perhaps too many!) different project concepts. You may have been unsure as to how they relate to each other. The challenge of this module has been to hold space for these concepts to speak to you and reveal their connections to each other. Some connections may be immediately apparent; others will take time to understand.

Writing an RD is the first step in making your project cohere narratively. Don't think of it as an end product but as a beginning foundation for project communication. Expecting and grasping for end products too early will not only create emotional overwhelm, but also block your ability to really see the project materialize on its own terms. Lean heavily into this process by tracking what you think you understand about project connections and where the holes and fuzziness show up.

Consider starting your group work with reviewing your agreements.

1. Talk about your relationships to process versus any desires to get results *now*.

2. Discuss how your 25-word and 100-to-150-word RDs are different. What do you notice about how the short and longer RDs relate to your Key Descriptor tables and concept maps? This really requires talking about how narration is very important for representing the broadest stakes of the project.

3. Read each other's RDs and use your Key Descriptor tables, and concept maps to help anchor your discussions. Ask about each other's work, including your own. Feedback can include the following:

    a. How well related are the project concepts in the RD? How well do the key descriptors hold the RD together? Expect both obvious and fuzzy connections.

    b. Does the RD do a good job of zooming in on the overarching research problem or does it spend a lot of time on smaller details?

c. Does it focus too much or too little on theory, the empirical, or concepts?

d. How well balanced are the 4Ps?

e. Was it difficult or easy to find multidimensional project possibilities? What was it like to try to figure those out in narrative form?

f. Compare your RD to the concept map. Which concepts and other project elements are represented in narrative form? Is anything left behind? Is there any place where there is too much focus? What feels fuzzy or missing? What feels like it's approaching consonance and balance?

4. If anyone is having trouble translating the Key Descriptors table into a 100- to 150-word narrative, consider collaborating to examine examples provided in this module. Discuss how the Key Descriptors tables compare with the corresponding RDS—how are these vital concepts accounted for in the composition of the RDS? Then, reverse-engineer one of the RD examples (in Example 5.2) by evaluating the purpose and strategy of each sentence. Also discuss how (and why) the sentences are logically ordered. Or you can use each other's work and do the same process.

5. Now that you have four categories filled in on your grid (4Ps, Broad Contexts, Key Literatures, and RD), check to see that there is consonance across categories. That is, do your literatures match up to your RD? Discuss in your group how your work is congruent (or not).

You will hopefully walk away from the group workspace with some inspiration. See if you can set aside a small amount of time to immediately follow up on any small adjustments you want to make to your RD or general concept work so far. Or at least make notes about adjustments you want to make in the next several days. These can include rewriting your research imaginary, revising your whole Key Concepts table, reconsidering your 4Ps and/or literatures, or revising your RD. Begin to *feel into* what revisions and reiterations are going to work before you start Module 6 (and think about whether pausing and letting go are needed instead).

MODULE 6

Perceive Your Multidimensional Object

Multidimensional project concepts are diverse and often nonlinearly related, but they hang together. Their connectedness is present in the dynamic *tension* of the crosscutting concept combo(s) as well as in the broadly scoped *integrity* of an effective research description. As we said in the introduction, multidimensional projects have *tensegrity*. But how can you maintain a sense of your project's unique tensegrity, especially as you move toward writing a full grant proposal?

You can perceive project tensegrity via a simple but powerful multidimensional object (MO). The MO is a two- or three-word phrase that answers the following question: "What connects the broadly diverse empirical and theoretical dimensions of this project?"

The following MOs, which we will elaborate on, answer that question: "liquid land," "security ecologies," "im/mobile memberships," "food sensing," "moral energy," "securing the family," "financial salvation," and "coding blight." Your MO will play a big role in the rest of your concept work. It provides an iterable touchstone for maintaining dynamic congruence between your research description (RD) and, as we detail in upcoming modules, your broad and data-focused research questions.

Craft a phrase that evokes the project's conceptual multidimensionality and structural tensegrity.

There is a relationship between the RD and MO, but they have different project design tasks. As you know, the RD's job is to enumerate the project's primary ethnographic and theoretical features and to convey its core multidimensionality and disciplinary value. In contrast, the MO gives you a sense of the project's fully integrated congruence, even for the concepts that aren't listed in the RD but are found in your Key Concepts table and concept map. As such, the MO is a *heuristic device* that helps you *feel in touch with all your disparate concepts*. With a concise and powerful language object like the MO, the tensegrity of your project's conceptual relationships is elegantly evoked.

Multidimensional object-like devices can be found in many creative works. In art and music, themes are MO-like. Ethnographers often feature MO-like thesis phrases in their finished works. Examples of MO-like phrases from well-known ethnographies include Kim Fortun's "enunciatory communities," Aihwa Ong's "flexible citizenship," Juno Parreñas's "decolonizing extinction," and Savannah Shange's "progressive dystopias," among many others.[1] We also have MOs in our own ethnographic projects. Kris Peterson centered her fieldwork on Nigerian marketplaces and global pharmaceutical supply chains around the capital and chemical dimensions of "drug circuits"; Valerie Olson's fieldwork on US spaceflight focused on the governmental production of "outer ecosystems" as extraterrestrial high grounds that define the political stakes of terrestrial environmental sciences, technologies, and cosmologies.[2] The MO that you use to design your project may continue to be useful in your postfield analysis processes, or it may not. You may change your project's MO after data collection and analysis and use that new one to design your postfield creative works.

As a device that manifests a congruent framework of inquiry, your MO bridges two major research design phases. It links the "what is this project about?" phase (Modules 1–5) with the "what am I asking about and why is it important?" phase (Modules 7–9). In this second phase, having an iterable MO specifically helps you to hold the project together. Making project elements connectable helps to express its multidimensionality in the overarching research question and project significance (Module 7) and in data-collection questions and strategies (Modules 8 and 9). After learning how to identify and iterate an MO, you can refine its multidimensional possibilities at any stage of your project.

As you proceed with the next modules, your key concepts or your MO—or both—may need more revision. Don't be concerned about this! We train you to go back and iterate project elements as needed. Those needed iterations can happen without your project falling completely apart; instead, it will gain strength and congruence. Our process is based on this foundational premise: *your project's tensegrity can withstand careful reorganization; and you can reiterate project features like your RD and MO without completely going back to the drawing board.* Doing MO/concept iteration work is effective, efficient, and rewarding. It will help you stay confident about the cross-cutting congruence of your framework of inquiry.

Keeping the project congruent is vital for proposal support and funding, and your MO helps you ensure that. As a simple anchor for complexity, the MO supports your efforts to communicate your project, whether the exact phrase makes it into a proposal or grant. It may stay in the background of your process and you may not use your MO when you talk about your project. Or it may become part of the title of your thesis, dissertation, or book. Either way, it will stand as an organizing element for any kind of research communication. You will continue working with this MO—and perhaps change it—during postfield analysis phases. Even when you move on to other projects, keeping the MOs of all your projects in mind can be a way to identify the overarching trajectory of your overall scholarly research program.

This is all to say that MOs do a lot of work. But, most importantly, an MO that *feels intuitively right* in your body/mind does that work effortlessly. Moreover, it provides a foundation for motivating yourself and exciting other people about your project.

The following exercises will help you perceive a serviceable MO over five days. We say "perceive" because the process is an act of feeling into your project's overarching projectscape, which can sit just below everyday levels of awareness. To get on the path, we invite you to take a creativity-boosting pause. Then we'll introduce you to some RDs and their attendant MOs and to the multidimensionality they open up. You will then draft and assess an MO. It involves returning to your concept map and assessing if your MO relates to, and creates relations between, all your key concepts—even those that aren't in your RD.

In between these exercises, it can be helpful to process your progress with others—either learning partners, such as students or friends, or mentors and advisers. Finding the MO that best conveys your project's tensegrity is an unusual quest. It can be difficult to pin it down on the first try, so we offer some assessment and reflection exercises at the end of the module. Upcoming modules will continue to refine this work. But it's a process that is deeply enhanced by sharing its ups and downs, and embodied joys, with others.

Exercises

EXERCISE 6A Revise Your Key Concepts Table, Concept Map, and RD

There is a possibility that some of your key concepts, or one of these concept-work elements, changed after you did group work in Module 4. If anything did get revised—or if you now think that something should be revised—make sure that you take time to make these changes in the Key Concepts table, concept map, and the RD so that they are congruent. The pause to reflect and revise is vital for the next steps.

EXERCISE 6B Pause to Feed Creativity

Here's a thought that is strange but true: your MO already exists! It exists before you see it on paper or on-screen. When you trust that it already exists, the only thing required to do is to perceive it. The key is to listen to your project's frequency, tune to it, to be open to what its connectivity wants to reveal to you. This practice of attuning to a project's conceptual and intentional congruence is a creative process. We hold that it's innate for people to feel into the dynamic relatedness of things, no matter how much societal interference knocks that reality out of our consciousness. Listening to your project connects it to the larger harmonic of creativity and possibility.

What gets in the way of perceiving for an MO to appear? Sometimes certain kinds of personal and communal experiences result in creative blockages. These may be expressions of social or personal wounds or traumas, often stemming from persisting imperial structures. These blockages and orderings affect folks differently. They can alienate us from the body, which is the ground where all connection, present moment awareness, and intuitive knowing reside. So leaning into creativity and intuitive knowing is about connecting to a politics of repatterning and course-correcting our inhabited subjectivities hailed by hierarchical, racial, gendered, and violent regimes. It's important to find ways to engage project listening on our own terms.

Your quest to perceive your MO doesn't have to feel like it's going to generate some momentous paradigm shift in anthropology—that's not the point. Just being with your project in an open listening mode can shift how you groove into a *conscious relationality* with your project, with anthropology, and with others in your worlds. Such processes of attunement can open rich avenues of creative thought and connection. This experience conjoins research design to what it means to craft and build a life's work.

Here are some exercises to help with your creativity, intuition, and project listening. As you try these, feel Miles Davis: "Do not fear mistakes. There are none."

- *Sit and be.* Spend ten to thirty minutes in a place (a park, a porch, a quiet room, etc.) practicing alone-but-in-worldly-communion as often as is possible during the week. When you set out to have that experience, do not bring anything to "do" per se. Just feel into where you are and listen to what speaks to you in that field of listening. After you're done, free-write about what happens sitting in stillness. Where does the mind go? Is the experience pleasant or difficult or both? Do your thoughts turn to the idea of doing something more productive or better? Note that mind chatter might be getting in the way of creativity and creative thought. What are you blocking and cultivating when you believe the mind's negative reactivity?

- *Do some liberating thought experiments regarding your project.* Fill in the blank:

  "If I didn't have to worry about doing it right, I would try..."

  "If I can be wrong, I would try..."

  "If I knew I couldn't fail, I would try..."

"If I felt like my efforts weren't going to be judged, I would try..."

"If I could embark on something just because it feels right, I would try..."

– *Attune your senses.* This exercise helps with both concept work and refining your participant observation skills once you get into the field. Holding an attitude of an easeful exploration can help open to expansive creativity. We recognize that some folks may have limited access to one or more senses. Choose which are right for you. They can be done inside or outside, and you should dedicate fifteen minutes to them:

* Listen to ten to twenty different sounds or touch ten to twenty textures. Notice how your mind responds to these sense-stimulating objects (is it distracted, is it calm and present, does it judge the activity?). Allow your engagements with these sensations to come and go, noting the subtlety or intensity of sound or feel, and the ways they change. Notice that each moment will never be the same as the next.

* With eyes open or closed, note the sensations that come to your body at rest. These can include the sensation on the body surface (the body contacting a chair, the hands touching each other, the feet on the floor, etc.) or sensations within the body (the breath actively breathing the body; physical sensations like tightness, pain, openness; energetic sensations, etc.). Notice that each moment will never be the same as the next.

* If you would like to, do the same practices for smell and taste.

Practice these exercises over a period of time and notice what happens to your sense of relaxation, openness, and your intuitive capabilities.

EXERCISE 6C **The RD-MO Relationship and**
Multidimensional Tensegrity

There are no strict rules for how to identify an MO, but each of the two or three concepts should consolidate the tensegrity—the conceptual harmonies and tensions—of the whole project. However, we offer a helpful formula to kick-start the journey of perceiving your MO.

Typically, at least one of the MO terms comes from among your concept combo(s); the full phrase needs to include both an ethnographic concept and a theoretical concept. Here is a way to see this as a formula:

MO = Ethnographic Concept(s) + Theoretical Concept(s)

Let's break this formula down. One of the MO terms should be a broad but specific ethnographic concept (or a variation of it) that's located among your 4Ps. We are pointing you toward a term that connects the specific social groups you are working with to the main sites, relationships, and materialities of the project. Another term should convey the project's more extensively theoretical or contextual dimensions. By this, we mean a term that represents the structuring processes or dynamics that are intellectually and socially pivotal to your project. Note that, at times, the boundary between ethnographic and theoretical concepts can be thin or nonexistent.

In the MO examples introduced at the beginning of this module, you will immediately note that each includes one or two of the concepts from the designer's multidimensional concept combos. To show how the terms in each phrase relate to our formula, we bold the broad ethnographic concept and italicize the more theoretical or broad contextual concept: "*liquid* **land**," "**security** *ecologies*," "*mobile* **memberships**," "**food** *sensing*," "*moral* **energy**," "*securing* **the family**," and "**financial** *salvation*."

You're probably saying to yourself, whoa, each of these concepts could be specifically ethnographic *or* broadly theoretical, depending on the ethnographer's project goals. This is true. In these examples, "security" serves as an ethnographic term in one project and a theoretical term in the other. But the important point is that, in either case, the combined concepts of the MO serve to evoke both the particular focus and the broad spirit of the project's features and aims.

We'll figure out what those terms might be for your project in subsequent exercises. But first it's important that you can see how an MO relates to concept combos and RDs. It's also important to understand how, in its flexible but overarching way, it represents the consolidation of diverse project concepts. In other words, we want you to perceive how it represents a project's unique multidimensionality as well as its overall conceptual landscape.

In the introduction, we elaborated on Tariq Rahman's "liquid land," Jason Palmer's "im/mobile memberships," and Kimberley D. McKinson's "security ecologies." In Example 6.1, we provide additional RDs with accompanying

EXAMPLE 6.1 The RD-MO Relationship and Multidimensional Tensegrity

RD	MO	Key Descriptors
Forest Haven Given ongoing conflicts between state authorities and Alaskan Native peoples, this project examines how twentieth-century assimilation policies still inform the state regulation of Indigenous food practices. It focuses specifically on the sensory dimensions of Native food gathering, hunting, and processing. It positions sensing as a site through which forms of knowing and sovereignty are negotiated and enacted. In this way, the project can examine food sensing at the level of everyday embodied Native food practices and the extreme technical practices of Alaskan state authorities regulating Native plants, animals, peoples, and practices. This project aims to analyze the persistent strategies of Native people to respond to shifts in the ongoing project of settler colonialism.	**Food** *sensing*	Indigenous food / Alaska / settler colonialism / sensing / Alaskan Native Peoples / state authorities / Indigenous practices / sensory regulation / assimilation / State-Indigenous conflicts / sovereignty / indigeneity / food sensing / everyday life / technical practices

Annie Wilkinson The movement against "gender ideology"—a concept used by right-wing groups globally to reference the social construction of gender—is gaining force across Europe and Latin America. Over the past decade, pro-family movements have claimed that gender is rooted in nature and that this scientific fact must be defended. This allows right-wing organizations in Mexico to claim that they must secure the family against gender equality and LGBTQ rights. In doing so, the Mexican pro-family movement makes "gender ideology" a national security problem that bears on state corruption, violence and crime, and neocolonialism. Using ethnographic research, the project follows how right-wing theories of the family connect to theories of the state. It also investigates how gender and security agendas are being shared among authoritarian populist movements across the globe.	*Securing* **the family**	Mexico/ right-wing movements / "gender ideology" / security / right wing activists / state authorities / LGBT people / Mexico / security policies / gender violence / global populism / human rights / securitization / profamily legislation / corruption / gender / equality/ sovereignty/ gender conflicts / regional authoritarian movements / the family
Ellen Kladky Within deindustrializing Appalachia, there is a growing movement to manage family life using religious financial practices. This project examines intertwined changes to financial subjectivity, family life, and social class through an ethnography of Christian financial literacy and savings programs in the region. Based in working-class communities in economic transition, it follows what happens to white	**Financial** *salvation*	US / Appalachia / families / Christianity / financial practices / financial subjectivity / deindustrialization / debt / finance / regional history / economic transition / livelihood / regional crises / new economic practices / class

identity and family processes as communities lose historical forms of livelihood, take on debt, and reorient to racial and class privilege. Situated in broader interconnected opioid, environmental, foster care, and job safety crises, the project examines how novel economic practices are changing the concept and experience of class in the United States.		
Nandita Badami India has become the second largest solar market in the world. The appearance of sunlight as an industrial energy form in India intersects with its historical presence as a religious power source. As a result, the politics of sunlight in the contemporary context of Indian nationalism is both physical and metaphysical. Situated in the outskirts and interior sections of Mumbai among policymakers, solar technicians, solar smugglers, off-grid builders, and spiritual centers associated with the sun, this project investigates how India's contemporary political conflicts are shaped by people's different engagements with sunlight's promising "alternative" characteristics as clean, free, and unlimited. By investigating Indian solar power as a contested moral and political force in villages and cities, the project illuminates the combined cultural and economic dimensions of national alternative energy transitions.	**Moral** *energy*	India / Mumbai / sunlight / solar power / energy / policymakers / solar technicians/ solar smugglers / off-grid builders/ solar technologies / solar spirituality / solar spiritual centers /solar markets / nationalism / clean energy transition / politics / morality / religion

MOS for you to reflect on. Remember the pairings and triplings you created in Module 4 when you were identifying multidimensional concept combos? Recall how you used your concept combos and key descriptors to compose your RD. Look at how the MO integrates all the key descriptor concepts. Feel into these MOS and their multidimensional tensegrity.

Each MO creates a conceptual snapshot of the project's focus and multidimensional dynamic. It bears repeating: the MO forms a flexible but overarching container for the RDS' ethnographic and theoretical concepts. As we will show you in later exercises, the MO can have concepts not located in the RD but are present in, or variations of ones in, the Key Concept table.

Any adjustment to those RD components *or* the MO has the potential to slightly or even dramatically change the entire meaning or conceptualization of the project. Holding the RD-MO dynamic in view allows you to figure out if there is a concept or conceptual relationship that's not working out in your project in some way. Refining the project is then easier with this tool that helps you keep a "feel for the project."

In the following examples, we narrate how this works for two of these RD-MO relationships. Specifically, we detail how these MOS connect to the 4Ps, broader contexts, and other social contexts that you began to establish in Module 1. We described Forest Haven's MO in the introduction and detailed her RD concept balance and multidimensionality in Module 4. However, in the following discussion, we assess the efficacy of her MO in relation to the RD and key descriptor concepts (see Example 6.2). We follow that

EXAMPLE 6.2 How the MO Maintains Multidimensional Tensegrity: Haven

Forest Haven's MO is *food sensing*. "Food" is the MO's ethnographic concept and "sensing" is the MO's theoretical concept.

Let's start with how it relates to the project's *4P empirical* dimensions. The RD makes clear the *people* are: Native peoples and state authorities. Food is a situated *part* that is socially *processed* and differently experienced through modes of embodied and technical "sensing" across *places*, which are being newly defined in Alaska in both historical territorial and technical environmental terms.

Let's look at the multidimensioning that "food sensing" creates for the combined study of food *processes*, political *processes*, and *Alaskan* places.

While "food" is typically linked with individual bodies in a metabolic way in anthropology, Haven's use of the term "sensing" covers more multidimensional ground than "the body." To be clear, "sensing" gestures to the political and social relevance of the right to relate to food as an intimate substance as well as to the production of settler colonial regulatory *processes* that control it. In this way, using the term *sensing* in the MO extends fieldwork possibilities from the most interpersonal and socially sovereign dimensions to the most spatially three-dimensional spaces.

How does "food sensing" link to *broader social contexts*? The project's RD contains several very important big-picture context project elements: the state, Indigenous food practices, settler colonialism, and food insecurity. As such, this MO is multidimensionally applicable beyond this situated case because it allows the project to bridge Alaska's unique context with other state, foodscape, sensorial, and settler colonial contexts.

In terms of *other key concepts* related to scholarly work, "food" and "sensing" are theoretical categories that invite examining how persons and institutions relate to food. The capacity to sense (taste, smell, touch, hear, feel) food is experienced through direct social encounters and mediated by tools and techniques. In this case, "food sensing" opens up a way to retheorize the state, indigeneity, and sovereignty.

Haven's RD-MO *relationship* has a particularly interesting capacity. The MO allows her to research how the state controls indigenous food procurement practices and eating in new ways, such as by making edible plants and animals into environmental governance objects related to managing industrial crises like commercial fisheries collapse and climate change. For example, she can do fieldwork on Indigenous activities to maintain sovereign sensory experiences in relation to the state's use of mechanical remote sensing technologies to police spaces and activities.

Her project instantly gains multiple dimensions as it goes from being a foodways ethnography to being also an ethnography of peoples' responses to the evolution of the settler colonial environmental state and its new ecological control schemas. Moreover, her project multidimensions the anthropology of food, the anthropology of Alaska, and Native studies with the literatures of political and legal anthropology, economic anthropology, environmental anthropology, and science and technology studies.

with a similar discussion of Nandita Badami's MO (see Example 6.3). Perceiving project connections across elements are vital to the process of effective iteration and multidimensioning.

EXAMPLE 6.3 How the MO Maintains Multidimensional
Tensegrity: Badami

Nandita Badami's example MO is *moral energy*. It relates the project's 4 *empirical* concepts at the national and transnational level, by following these *people*: religious practitioners, bureaucrats, solar technicians, solar smugglers, residents of solarized homes, and off-grid builders making Mumbai's solar *places*. The *parts* of sunlight, solar, the sun, and power come together in energy, which combines political and technical social realms.

"Moral energy" allows Badami to track the role of solar power as a compellingly "good" resource that acts as a symbolically loaded *part* to conjoin religious, political, technical, and economic *processes*. It turns "the sun" into both a part and a process of Indian life. Since "energy" is a term that refers to forces that are both material and sacred in contemporary India, the MO allows Badami to follow both the technical and religious logics and narratives that make solar energy futuristic fuel.

In terms of the *broader contextual* dimensions of this MO, "moral energy" exceeds the typical earthly economic and political realms, extending into the solar systemic dimensions of outer space-based technologies, into the temporal dimensions of Hindu religious ages, and into the spatial dimensions of Hindu cosmological dynamics. It also links the micro and nano scales of energy with the broadest contexts of physics as a social and natural force.

The *other key concepts* that this MO gathers together include nationalism, capitalism, democracy, and alterity as well as metaphysics, religion, transition, and the emerging concept of "alternative" power. The RD-MO relationship that "moral energy" sets up allows Badami to connect solar energy development with broader social and political *processes* aiming to power an Indian state that is "proper" to contemporary authoritative discourses of independence, homogeneity, progress, incorruptibility, law-enforcing, and righteous governance. Badami's project goes from being an ethnography of solar energy per se, to an ethnography of solar power, where "power" recognizes the multidimensionality of physical energy and powers-that-be. Her project can therefore multidimension the anthropology of India, science and technology studies, political anthropology, and the anthropology of religion.

As you can see, the MO makes connections across all Key Descriptors table categories. Take time to understand this process of articulating the ways an MO can create tensegrity among crosscutting concepts and hold a project's conceptual framework together. You'll write a short reflective memo after you land on an MO, as a way of testing your MO's potential and communicating its value to your group members.

Before you move to the process of perceiving your MO, take a moment to warm up. Free-write a response to this question, following the prompt:

1. *What will your MO's tensegrity tasks be?* You can free-write or voice-memo a reflection on the concept work you've done to make strong relationships between your key literatures, your concept combos, and your key descriptors. Reflect on the connective capacity that your MO needs to demonstrate in order to hold your project's concepts together. Don't overthink this free-write: the purpose is to simply acknowledge the conceptual connective tissue that you must feel to perceive an MO.

EXERCISE 6D Perceive and Create Your MO

Now you're ready to perceive and unearth your MO—or at least your first iteration. Whatever MO you land on now will need adjustments. For inspiration, we provide Tariq Rahman's timeline of MO iterations over a six-month period (see Figure 6.1). This shows stages in the concept work. Notice that working with six different MO possibilities ultimately led to "liquid land."

The takeaway message is that MO work takes time! Allow that to be the case. The process also requires emotional and embodied space to do the required creative brainstorming exercises. Try to approach this work rested, open, and ready to relax into the exploration. This exercise takes three days to complete, so make sure you plan for that.

Day 1. Gather Materials and Make a Space for the MO

First, you'll generate three candidate MOs, one at a time, by feeling into your RD, literature Venn diagram (with populated concepts) from Module 4, and concept map. This is a stepwise process, so stick to it the best you can.

1. *Make a space in your concept map.* Place your RD, literature Venn diagram, and concept map in front of you, perhaps at-

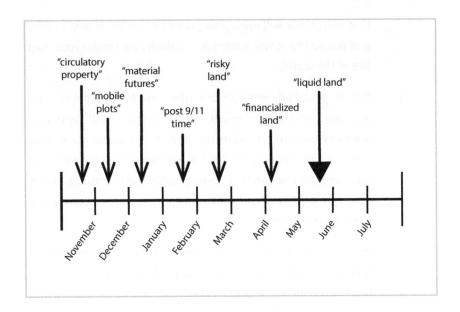

FIGURE 6.1 Rahman's MO Evolution

tached to the wall. This first step makes primary use of the concept map. Create a small blank space in the "conceptual middle" of your map. By "conceptual middle," we mean a spot on your map where it is clear that the phrase connects all the elements. This could be the middle, side, top, center, or bottom of the page, depending on your concept-map shape. For example, if you made an analog map, redraw to create a space for the "conceptual middle." If you used sticky notes to build a map, move them to create that "conceptual middle." If you created a digital concept map, move the terms outward to create a "conceptual middle."

2.   *See your map as a set of relations.* Meditate for a bit on your map and feel into all the concepts and the various conceptual linkages you have drawn. Get a feel for your map's "conceptual middle" space. Reflect on your concept-combo validation process and consider how its placement in the middle of the Venn diagram reflects the literature tensegrity in your project. Know that there is a phrase to place in the middle of your concept map that will feel as if it connects all these key project elements. Also know

that this phrase will open your project to worlds of new inquiry and possibility. It will guide your curiosity and make your map live in the world.

3. *Prepare your body-mind for brainstorming.* Now comes the creative turn, so take a moment, like any artist or athlete, to prepare. Review your RD and recognize its construction out of your key descriptive concepts. Check in with your Venn diagram map and see how the concept combo holds the literatures in creative tension. Now, hold your concept map and its "conceptual middle" in full view.

4. *Brainstorm an MO phrase that contains at least one empirical concept and at least one broader theoretical concept.* The best two- to three-word phrase should feel like it fits in that "middle" space. It should have at least one of your concept-combo terms. Your key descriptors in the RD are also valid sources. But do consider your full map for other possibilities.

Following the spirit of Example 6.4, write your first candidate MO on a sticky note and place it in the middle of your map with the open center. Take an hour-long break.

EXAMPLE 6.4 Rahman's MO Perception

> Remember how, for Rahman, in the end of his MO perception journey, "land" was the clear ethnographic concept that united all the empirical dimensions of his project and was key to his concept-combo literature searches? But what was the quality of this land that his focus on financialization was surfacing? After several MO tries, a surprise emerged: in his research imaginary Rahman had noted the liquid feel of his project *and* of land in Lahore, and he had included "liquid capital" in his Key Concepts table. Therefore, after several tries at landing on an MO, "liquid" arose as the right theoretical tensegrity concept when he felt into the emergent financial and technical changeability and capitalized flows of land in Lahore.

5. *Open the space again.* Remove your first MO (on its handmade or digital sticky note) from the map. Take an hour or longer break.

6. *Repeat!* Start over, redoing steps 1 through 4 two more times. You've now produced an "MO short list."

Day 2. Compare the Three Candidate MOs
in Relation to Your RD

Take a moment to simply play with the three MO candidates—to just do nothing other than to observe and listen to them. Allow yourself to just simply be with them without any agenda. Open up to how each feels in the body, without any judgment or shoulds or efforts to try to be "right." Do so by following these prompts:

1. *Make an MO testing zone.* Place a copy of your RD on the wall. Make three copies of your concept map with open "conceptual middles" with each containing one of your candidate MOs. Place each map under your RD.

2. *Feel into the connective potentials of each MO.* For a day, find some moments to sit and meditate on this assemblage of your RD and three MO candidates. Let your mind shift across each concept map from center to outside, and then outside to center, testing your MO as a connector. Let your mind shift between your RD and each concept map, between narrative writing and images of terms, testing each MO as a potential bridge. As you do this, you may even come up with an MO that feels better than your three candidates.

3. *Note what you experience.* Write freely to yourself about how you experienced this process. Or you may discuss with someone else and make notes of what surfaces.

Next, you'll conduct a more formal assessment.

Day 3. Analytically Reflect and Select Your MO

It's time to formally reflect on your three candidate MOs and land on the best one to work with. Here are some prompts to answer for each one. Refer to our analyses of the example MOs above to help you do this. Remember that, just like the RD, the MO must be balanced. In this case, balance means that the MO connects as much to your 4P elements as it does to your big-picture context and other key concepts. By "connect," we mean that you can perceive and imagine narrating that connection, even if it is roundabout and general.

1. *Does the MO communicate congruence and the spirit of your framework of inquiry?* Consider whether the MO creates connective congruence within your map and invokes the general aims of your RD. To test this, fix your eyes on each term on your map and think about how each connects to the MO. Do the same with the sentences of your RD. Write about which concepts/phrases seem to connect best and what seems to connect less well. Here are some specific prompts to answer, based on the evaluations we wrote for the Haven and Badami MOs:

    a. *Does the MO draw most if not all of your 4P concepts together?* We simply want you to be aware of when a 4P concept seems to fall a bit outside the MO's umbrella. This may be, for example, because that particular 4P concept is closely related to another more key 4P, such as one that shows up in your Key Descriptors table. In that case, the seeming disconnection isn't something that would call your MO into question. However, if there is a more significant disconnection between a 4P concept and your MO, take a moment to address it; ask what it might mean for the congruence of your project elements and your selection of an MO.

    b. *Does the MO connect your broader context and other key (theoretical) concepts together?* Does the MO seem to link each of the broader contexts you identify for your project and also seem to illuminate their interconnectedness?

Does the MO's theoretical concept connect to the kinds of literature-based concepts that you imagine contributing new theory to?

c. *Does the MO link to the RD to invoke an exciting fieldwork space?* Does it connect well with all the processes and ethnographic engagements suggested by your RD, and does it help show how it will bring together and contribute to your literatures?

2. It's time to imagine the MO's multidimensional possibilities. Address the following prompts:

a. *Can you see multidimensionality emerging with this MO?* Look at the RD and also at the multidimensional concept combo you landed on, and validated, in Module 5. Look back at the Key Descriptors table you developed in Exercise 5d. How does your MO represent the combinations of empirical and theoretical elements that you developed there to highlight your project's multidimensional value? When you zoom out to all concepts in your map, and the lines you drew between them, how does this MO help you see how all the concepts matter and belong to your project, even if they are not linearly or directly related?

b. *Does it invoke the necessary empirical limits?* How is this MO *bounded* enough to hold together the specific empirical focus and internal congruence of your project?

c. *Does it evoke enough theoretical scope and broadness?* How does this MO *open* to intuitive hunches you are now having about how this project is related to broader contexts? Or that are part of other projects you've read about?

3. After responding to all the previous prompts for your MO candidates, now you will select the most promising and dynamically connective MO. Here are some "if-then" prompts to help you winnow out your best MO candidate.

a. *If a strong MO emerged....* If, for any of the MOS you chose, the answers to all the questions seem to yield a palpable "yes" in your mind-body, then good! You have an MO candidate. Hold on to it and expect that changes may likely occur with more concept work. Indeed, the MO can change as more is learned and understood about the research project.

b. *If you have one or more "almost right" MOS....* If any of your MOS produced answers that were "yes, but ...," then consider for now that these may not be the right MO. If you intuitively feel that one of these MOS still might work, despite its shakiness, then name how/if the connections you're drawing together work or don't work—the process of naming can help clarify fuzziness. It may be that iterating some of your previous concept work will bring it into focus.

c. *If you felt some strong nos....* If any or all of the answers to these questions for one of your MO choices was "no," then the chances are this isn't the right MO at this stage.

d. *If you have two MOS that seem equally good....* If you identified two possible MOS and then put them into deeper analysis, you need to land on one. When you do, don't worry that you have to ignore the other candidate. Try this: can they be hybridized to make one better MO? If that didn't work, simply note your other MO and discuss both when you get into your concept workspace.

e. *If you didn't feel good about any of them....* If, after all of this, you feel that you did not figure out one solid MO, take heart! Stuckness is productive. There are many ways you can "unstick" yourself. You can try remapping your overarching concepts, discussing with a friend, or taking a quiet walk alone. This may be enough to help you find the best MO; if it doesn't, continue with the following exercises.

If, after you run through these prompts, you found a strong MO, stop and go to Exercise 6e.

4. *Get unstuck if you're stuck.*

   a. If you can't unstick and find your MO, we recommend taking a pause. You may simply be tired or unable to "see the forest for the trees, or the trees for the forest." This is a great "flippable" metaphor that helpfully sums up the concept work that guides your project's specificity and generality to be balanced so you can find a strong MO. Take a day to rest and relax.

   b. After you rest, follow these free-write prompts to assess how to proceed:

      i. *If you are stuck, where are you stuck or what does that stuckness feel like?* Can you move into stillness and ask the stuckness to talk to you? This conversation might be important because it's likely something you don't expect.

      ii. *Is this process causing you to rethink the RD and its components?* How are you feeling about the basic description of your project? Does it feel right or do you feel "pulled" toward another way of imagining your project? If the latter, you may need to revisit Module 4 and see if your RD needs work. Are you feeling alone and confused? Consider proceeding further with some help from your community of designing colleagues.

      iii. *Reflect on your answers to these questions.* Your answers will tell you a lot about where you are at and what you need to do. If the first question helped you relax and be open to the process, you simply needed a break and to acknowledge some anxiety or pressure. If the second question rang true, this is the time to consider which of the modules or exercises you may need to repeat to get to this point.

Don't let this fact make you feel exhausted or discouraged! Investing in repetitive iteration right now will save you lots of time later.

EXERCISE 6E Feel into Your MO

Now that you've gone through all MO steps and reflections, how does it *feel*? Feel into your bodily reactions and thoughts. Spend about fifteen to thirty minutes free-writing about your MO. Here are some prompts to consider:

- How does it feel in the body?

- What do you intuitively feel might need to be tweaked about this MO as you move forward in the next modules?

- What creative insights do you have about your project now?

- What feels exciting and inspirational about your project?

- How can you see how your project connects to those of other creative works that you admire?

Now you'll be ready to put your MO in your grid and share it with others in your group. Celebrate this pivotal moment—you and your groupmates have perceived how each of your projects hangs together.

EXERCISE 6F Research Project Grid 6

Place your best MO in the Research Project Grid table (even if it's not feeling or landing right). Up until this module, you have been using the grid boxes largely as important placeholders for concepts and ideas. Now that you have developed your MO, future concept work will also be used to assess congruence across all the grid's categories. When you change any aspect of your project, make the change in the grid. Adjust the related elements accordingly so that you retain integration and congruence. In other words, if a literature category changes, then all categories need to be adjusted along with it.

Collective Concept Workspace 6

You are now halfway through the modules, and this one was probably the most challenging. Know also that you are perceiving your first of most likely many MOs to come. Recognize the hard and committed work you've completed so far.

Before you meet, share this module's concept work with your group members. Come prepared to present your work:

a.   Make and bring a document (or a small slideshow of three slides) that contains your RD, your concept map, and your MO.

b.   Be ready to read your RD, explain your concept map, and then introduce and explain your MO selection. Your explanation should justify your choices along these axes: ethnographic, contextual, conceptual, multidimensional.

As you begin your group work, consider acknowledging and holding space for a sense of ease, self-compassion, and compassion for others. This is very helpful for settling into the group process. Lean into the process over and above expectations for a solid outcome; keep cultivating an attitude of friendliness toward your research project. Before your group begins, consider revisiting the agreements.

1.   Have each group member present their RD/concept map/MO document for ten minutes. Address these questions: What was your short list for the MO? And what did you decide on and why?

2.   How does the MO hold together all the key descriptors in the RD? Reflect on what connects and what doesn't.

3.   After you have explained your MO process to your group, talk about whether it's feeling "right." This is an intuitive knowing and often people feel the "rightness" of the MO in the body. Excitement, elation, and a sense of lightness are always good signs that the MO is landing well in relation to the RD (even though it may change as you get to know your project and build on the existing project elements). Another sign that it's landing well is that the entire group also feels a corporeal resonance of the

MO. If you're not feeling a bodily resonance like this or if you're feeling like something is "off," that's okay and actually quite expected. Work with your group to figure out which exercises (or maybe even free-writes or further discussions) you might try to get a breakthrough. Always consider taking a break (and take seriously/identify what that means for you) for a rejuvenated and refreshed perspective. Continue to come back to it. For now, allow the current MO you've perceived to be "good enough" for the next and upcoming modules.

The Inquiry Zones

From the very beginning of this journey, you've had a research idea in mind. You want to understand something about the relationship between people's lived experiences and larger forces. What underlies your interest is a spark of curiosity motivated by scholarly, creative, political, social, and ethical interest. Your questions so far have been broad and open: What is this new sociocultural process or problem? How do people experience it? What does it mean for how people relate and live? How might the world be otherwise if people understood the sources and entailments of this process in fuller ways? Such questions and concerns animate the heart and soul of ethnographic research.

INTERLUDE 2 PURPOSE
Learn about project inquiry zones and how to integrate them.

Typically ethnographers try to settle on research project questions early in project design. But we take a different approach. We find that it's best to hold the questions in an inchoate space of interest until you develop the project's descriptive field of elements and connections—as you have done so far. The questions that emerge out of those efforts tend to be strong, congruently integrated, and deeply significant.

You have completed all the concept work that you need in order to begin designing your research questions. You will develop those questions within three "zones of inquiry": scoping, connecting, and interacting. The *scoping zone* is where you *scope out* the project's broader significance and disciplinary and social space, as well as its overarching research question. The *connecting zone* is where you group project elements into interrelated thematic and process clusters so that you can *connect* that overarching question with three to five specific data-collection questions. The *interacting zone* is where you generate questions for your project's interrelational fieldwork processes: interviews, participation, archival research, etc. As you can see, all your project's questions are directly linked to each other within a multidimensional framework of inquiry.

Throughout the three inquiry zones, we distinguish between the purpose of each zone's questions. Typically a scoping question, data-gathering question, and field-based question are all generally, and unfortunately, referred to as "research questions." This is confusing to first-time research designers who don't understand how questions possess different forms and functions. In the following modules, we make these distinctions clear and highlight the project design pitfalls in not understanding them.

We use the term *zone* to connote an embodied workspace as well as a state of mind. Moreover, "in the zone" is a colloquial phrase in the United States that signifies the attentional congruence of body and mind within a state of positively elevated emotions that connect with others. These inquiry zones are more akin to the ecological, not geographical, spaces of living relations. Each zone has its own unique concept-work focus, specific kind of research question, dimensioning possibilities, and political and ethical intention. But they are not hierarchical—far from it. Like the domains of fauna and organisms in the contemporary biological imagination, the zones are interdependent. You'll discover attending to *any one of these zones can be used to reshape the others*. As with any of the processes in this handbook, we provide a basic order to begin working in these zones, but you can and should work iteratively between them as your project enacts in the world.

Each of the zones has three design dimensions: *field, multidimensional,* and *ethical*. The field dimension contains the project's general sociopolitical and conceptual field as well as its specific fieldwork space. Connected to this field of thought and work, the multidimensional *opens* possibilities for creative connection, inquiry, and analysis that, as we describe below, support strong project significance. In addition, each zone has a relational dimension of politics and ethics. The scoping zone consolidates the politics and ethics of its broader disciplinary and social commitments. The connecting zone articulates those politics and ethics in terms of research organization, and the interactive zone is where you enact those commitments. Table Inter.1 provides some definitions.

TABLE INTERLUDE.1 Project Zone Dimensions

Zone/Role	Field Extents	Multidimensional Possibilities	Ethical Milieus
Scoping	Overarching conceptual and social field *Field inquiry frame*: overarching research question–the "scoping question"	Opens the project's disciplinary, 4P, and social/contextual extents of inquiry; establishes overall significance	Establishes the most extensive political and ethical stakes
Connecting	Multidimensional project articulations *Field inquiry frame*: data-gathering questions	Opens multidimensional connective possibilities of data collection	Articulates political and ethical intentions
Interacting	The space of fieldwork *Field inquiry frame*: embodied field-working questions	Opens the possibilities of interpersonal inquiry and interaction with people, parts, processes, and places	Enacts political and ethical commitments

It may seem counterintuitive to begin with the overarching scoping question rather than the data-collection questions or the face-to-face inter-acting questions. But it's important to begin with the scoping zone because it maps out the project's big-picture anthropological aggregates—such as com-munities of people, shared kinds of experiences, types of spaces, and varieties of practice and process. The scoping zone is where the broader theoretical and social interventions of the project reside. These big-picture factors *di-rectly inform* how the specifics of other project design zones are fleshed out and interrelated. They also directly convey the field research strategies and necessities that answer an overarching scoping question pertinent to the overall aims of the project.

The dynamic multidimensional linkages you cultivate among all three inquiry zones both focus and expand the project's tensegrity—meaning its capacity to hold equally important but disparate elements, concepts, and aims together. You may, for example, in the connection zone, realize that you need to collect data on a process you haven't accounted for yet. Or you may, in the interactive zone, have revelations about who you can interact with and what you can participate in socially that make you rethink your scoping zone and question. We provide prompts for you to iterate previous work in order to account for these additions, before you circle back to iterate the questions you are posing in each of the zones.

Asking questions moves your project into the final phase of becoming, as we promised from the beginning, a multidimensional framework of inquiry fleshed out with imaginative and intuitive contours and open to otherwise sensibilities and possibilities.

MODULE 7

The Scoping Zone

What is your project's main research question? What value is there in answering it? You'll compose that question and begin articulating project significance here in the scoping zone. "Scoping" a project delineates its conceptual breadth as well as its limits. You need to clarify both to compose an overarching research question and to specify the project's overall intellectual and social significance. By using the term *scope*, we evoke the colloquial idea of "scoping something out," which is a way to be curious about something's broad potential.

Drawing on previous concept work, you will compose a *scoping question* that overarches your whole framework of inquiry. It will provide a strong but iterable foundation for subsequent sets of project questions: *data-collection questions* (Module 8) and *interacting questions* (Module 9). As mentioned in Interlude 2, these distinctive kinds of research questions must be carefully crafted to work together.

The process of composing this broad scoping question is, paradoxically, another exercise in project contraction. This question holds space for all your specific ethnographic, contextual, and disciplinary concepts. The strength of a scoping question is that it cannot be answered in a few paragraphs or by collecting one specific kind of information. It is a question that, when answered, will produce a book, many articles, a film, a new scholarly program, or all of the above. The scoping question also conveys the significant impact of a project's final outputs—for disciplines and particular social groups as well as for new ways of perceiving worlds and enacting futures. This is why the multidimensional design work of inquiry scoping and stating significance is in the same module.

MODULE PURPOSE

Scope out the project's overarching research question and significance.

All researchers struggle to produce an overarching question that is both groundedly specific enough *and* theoretically broad enough to justify a project to supporters and reviewers. Creating that question's broad-but-specific balance is key to maintaining project element congruence. As we've said often, projects read as well designed when they demonstrate balance and congruence. When researchers can't figure out an overarching question that is cogent but fully expanded, they often end up posing a half-baked question that doesn't clearly convey the project's comprehensive scope. This is why we bring you to the research question composition process after a lot of foundational concept work. That way, you won't try to compose a scoping question too early or too late. Furthermore, you will keep revisiting your scoping question as you do interactive work in the other inquiry zones. At the end of the handbook, you will have clear and dynamic congruence among all the questions you need to pose to make your project valuable and achievable.

In our experience, people often get stuck with an underscoped question because they don't understand the difference between a *main overarching* question and *data-gathering* question. While the latter directs you to gather specific kinds of data (Module 8), the scoping question is a theoretically inclined question that requires data collection as well as data analysis to answer. Therefore, it must be posed at an empirical and conceptual dimension much broader than the connecting zone's data-focused questions.

Because the scoping question overarches your project's data-collection processes and presages your final analytic contribution, working in the scoping zone helps you identify the overall significance of your framework of inquiry. A project can be significant because it researches, for example, a new identity, category, or process. But that kind of contribution alone usually does not have the weight to warrant support or funding. A compelling project significance statement conveys its disciplinary value and conveys what would be lost to scholarly and social worlds if your project was never done.

You may have noted our repeated use of the term *overarching*, as in "overarching research question" and "overarching significance." The metaphor of the arch will help you to imagine the purposeful shape of the scoping zone. An arch brings to mind a broad but elegantly delimited contour: the arc of a branch or natural bridge, the arch of a roof, or the arc of a story. Every project must be "open-arched" enough to cover broad connections and possibilities, but anchored to the ground in a way that creates necessary and practical research limits. This arch helps you contain the conceptual openness that allows you to explore the complexity of real-life interconnections within reasonable bounds. Those reasonable bounds reflect your goals and your fieldwork parameters. As with any social process, clearly stated boundaries are as necessary as openness.

The project overarch, with its broad openness and specific boundaries, comprises the spirit and work of the scoping zone. Clearly delineating a project's partially open, partially closed scope also steers you away from defining its conceptual and empirical range by using only totalizing or artificial concepts (like *development* or *neoliberalism*) or only vertical scaling dichotomies (like *local/global*). Understood this way, a project's *scope* doesn't have to be enclosed in an abstract, hierarchical order relevant to positivist social scientific discourse. Working with scope as a partially open, partially closed form that is broadly specific to the unique social worlds you will be in helps you get out of Western orderings and into otherwise conceptual and analytical terrain. This can inspire ways to organize anthropological work in terms of (currently) subjugated knowledges and experiences of connection and relatedness.

In the exercises, we walk you through several scoping-question development processes based on designer examples, beginning with Forest Haven's project. To introduce you to a strong scoping question and its value, let's consider how hers reflects her project's overarching scope: "How do conflicts over Alaska Indigenous food as sensory material reveal continuities and changes in the ongoing project of settler colonialism?" This question indicates that there is an emergent process that needs to be researched: new dimensions of foodway regulation. But it does more than this. It reflects the project's multidimensional connections, namely the link between the deeply embodied sensorial dimensions of Indigenous foodways and the broad regulatory and technological sensing dimensions of state control. It also gestures to the project's multidimensional tensegrity, which is held in the *food sensing* multidimensional object (MO).

True to the scoping zone's arcing form, this kind of empirically specific, but also broadly contextual, question both defines *and* limits the project's ethnographic activities. It opens a space of research but also acknowledges that a lone researcher, or one working within a particular research team, can only do so much, go so far, and relate to so many people! Therefore, a question like this frames the most comprehensive inquiry possible while conveying a sense of the project's ethnographically specific social and spatial workspace.

We provide lots of examples as we guide you toward a clearer view of your project's scoping zone, significance, and scoping question. And, like every other module, this one is designed so that you can return to it.

Before we begin, keep in mind that discerning the scoping question and project significance can take time. Remember that the work you produce reflects everything you know about your project at this moment. You will keep learning more. As you learn, you will develop new insights that will help adjust and deepen your concept work. So if you get stuck, it's important to know when to pause and when to keep going, which helps gain space and perspective.

You will begin the process of defining your scoping question by finding *scoping terms* among your key concepts. This is what we call the small set of key concepts that define the terms—the broad limits—of your project. Generating these terms, and your scoping question, will allow you to understand the social and intellectual significance of your project's particular scope.

To prepare for the exercises, gather your concept work from Module 6: your Key Concepts table, most recent research description (RD), and current working MO. It's also good to prepare *yourself*. Do this by ensuring that you have made time and space for opening up your conceptual bandwidth for new concept work. This means enacting self-care as best you can, whatever that means to you. If that's challenging, then try to create some intention in your schedule to support this open frame of mind. If you are especially having difficulty gaining relaxed mind states, you may try to limit hyperstimulating activities. Instead, take breaks, go on walks, pause in any way that resources you, and/or do short meditations focusing on breath and body. These practices all go a long way to helping you do the paradoxical work of holding the overarching dimensions of your work in your mind. This helps to contract your focus and write a couple of sentences that pose a broad question and relay the project's significance.

Exercises

EXERCISE 7A **Prepare to Define Your Scope**

Defining your overall scope involves a conceptually powerful act: identifying the *generally broad* ethnographic, theoretical, and contextual terms that are *specific* to your project. Based on your concept work thus far, these could be concepts like families, dancers, farmers, neoliberalism, digitization, food, money, land, labor, trees, the state, and laws. Such broad terms are made *specific* when linked to your sites (Alaska, Mexico, India) or to certain forms (Indigenous law, seaweed, venison, pesos, birch trees, ejidos, solar energy); they are also applicable to other social sites or processes. Such scoping terms provide a foundation for project resonance with other ethnographic work.

In this exercise, you will derive a succinct set of scoping terms from your key concepts. You will identify scoping terms by following a set of instructions. Because you are revisiting your RD to do this work, it is also a moment to recheck that the right broad ethnographic and conceptual terms are visible in that description. We established this in your previous concept work, but now is the time to confirm again.

In the introduction to this module, we revealed a scoping question for Forest Haven's project, and now is the time to show how this question can be generated. Example 7.1 shows what a finished Scoping table for Haven's project might look like.

EXAMPLE 7.1 Haven's Scoping Table

MO	*Food sensing*
RD	Given ongoing conflicts between state authorities and Alaskan Native peoples, this project examines how twentieth-century assimilation policies still inform the state regulation of Indigenous food practices. It focuses specifically on the sensory dimensions of Native food gathering, hunting,

	and processing. It positions sensing as a site through which forms of knowing and sovereignty are negotiated and enacted. In this way, the project can examine food sensing at the level of everyday embodied Native food practices and the extreme technical practices of Alaskan state authorities regulating Native plants, animals, peoples, and practices. This project aims to analyze the persistent strategies of Native people to respond to shifts in the ongoing project of settler colonialism.
Key Concepts	4Ps: *Persons*: Alaskan Native Peoples, Alaskan state regulators and authorities *Parts*: gathered foods, hunted foods, licenses *Processes*: Native subsistence food practices, state regulation of subsistence practices, food procurement and processing, sensing *Places*: Alaska, Native Alaskan communities, state offices Broad Context: sensory regulation and control, historic and contemporary Indigenous-state tensions, ongoing assimilation efforts of the Alaskan settler colonial state, climate change Other Key Concepts: embodiment, sovereignty, indigeneity
Scoping Terms	Alaskan settler state, Native Alaskans, regulators, food, sensing, settler colonialism

You will see the project's most "scoped-out" terms, which are distilled from the MO, RD, and Key Concepts table. Consider how the scoping terms link the project's particular empirical ground (Alaska, Indigenous peoples, settlers, regulatory processes, foodstuffs, food-producing beings) *with* broader context and conceptual domains (foodways, social governance, environmental management, climate change) *with* a form of perception (sensing) that links embodiment and technologies with historical and emerging domains of power (settler colonialism, sensory politics, resistance and refusal). Composing a question with these terms will contribute to the anthropology of Alaska and North America, Indigenous studies, the anthropology of food, and the anthropology of the state.

Importantly, these terms must reflect conceptual congruence among the MO, RD, and Key Concepts table. Let's consider what might have happened if Haven's RD and Key Concepts table weren't aligned. What if Haven's project had *the Pacific Northwest* in the RD but the Key Concepts table only contained *Alaska*, or vice versa? This is the moment to delineate which spatial limit the project is concerned with—the state of Alaska or the region of the Northwest? Deciding this has bearings on what counts as the ethnographical and theoretical limit of the project's scope—which is the specific state of Alaska in which particular Native peoples live and work with food and "the state" as a demarcator of political authority and a site that houses food-regulating authorities that will be interlocutors. To continue with the assessment, if Haven's Key Concepts table contained the terms *colonial* or *settler* but her RD did not, or vice versa, this would also be a moment to assess the broadest, processural theoretical scope (*settler colonialism* versus other legal or governing processes) that the project is concerned with.

Now comes the time for you to select your own broad-but-specific scoping terms. Here are some instructions to work through.

1. *Check for overarching conceptual congruence.* As per Example 7.1, check to ensure that your RD and Key Concepts table are congruent at the level of broad scope. If they are not, consider revising either. If you find that you need to do significant revision, then take a day away from your concept work before you move on to the next set of exercises.

2. *Select your scoping terms.* A good number of scoping terms to work with is four to six: two to three that are broadly ethnographic and two to three that are broadly theoretical or contextual. When you select a term, review the other terms to make sure this term is truly "overarching" in that it "contains" or hails other terms in your Key Concepts table. Moreover, check that it also gestures to broad connections between your project and other projects. Make sure also that the scoping terms reflect broad meaningful aspects of the MO's multidimensional concepts.

3. *Fill out your Scoping table up to the "scoping question" box.* Fill in the MO, RD, and Key Concepts boxes and then place the best four to six scoping terms in the indicated rows. Note that there is no "right" way to fill out this table (see Table 7.1), but

TABLE 7.1 Scoping Table

MO + RD	
Key Concepts	*People:* *Parts:* *Processes:* *Places:* Big picture context elements: Other Key concepts:
Scoping Terms	
Scoping Question	

you should be able to feel resonance between the broad terms we show in the Haven example and your own project's broadest terms. As you work, listen to your project, your body, and your intuitive knowing. Experiencing a-ha! moments while listening in this way usually provides good guidance. In the table appreciate the place where your scoping question will eventually sit. Begin to feel the kind of question this blank space can open up.

4. *Reflect on your project's "overarching" relations.* Take a moment to think about the connections that your project is making among you, others, and other creative works. Now is the time to think about group-work colleagues, your advisers, other researchers who work on this topic, the communities you belong to and hope to support, and new colleagues you hope to converse with in the future. Do your scoping terms indicate broad limits but broad openings between your project and others' experiences and work? Between your project and otherwise possibilities? Make any changes you feel are necessary to "specify" but also "open" the broadness of your project.

Now that you've done this work and reflected, do you see/feel the glimmer of your project's scoping question assembling into view? We'll get to that soon, but before we do, take a moment to pause and get into the mode of inquiry. In other words, feel into your deep curiosity, honor otherwise ways to inquire, and stand in the generative place of not-knowing.

EXERCISE 7B **Attune to Curiosity and Not-Knowing**

Over the years, we've noticed that students in particular find it challenging to ask questions for which they have no immediate answers. Enculturated in graduate school to "argue" or "state their thesis," they find it hard to shift into a genuine mode of open inquiry. But this can be difficult for any researcher at any stage of a career or project. This is why we establish *scope* as a kind of container for productive uncertainty. It is exactly your lack of certainty, within an empirical and conceptually defined frame, that leads to a good scoping question!

As discussed earlier in the module, Haven had an experientially informed intuition that her seemingly noncommensurate project elements—sensing, Indigenous sovereignty, the state, and food—belonged together in a framework of inquiry. And this is where the scoping question could be found.

Researchers who land on multidimensional projects with good overarching questions and significance do not, at this stage, know exactly what will be illuminated when they relate key concepts, but they take the relationships seriously. To draft a good question, they must not know! And so they often read extensively, write daily, engage in collective discussions, and sit patiently with uncertainties. You've been doing this too, and like them, you are probably feeling questions begin to arise that are both general to your theoretical aims and particular to your ethnographic sites.

Researchers can work with uncertainty when they stay curious and open, but it's not easy to stay there, especially in academia. Sometimes this uncertainty leads to fears that putting elements together won't lead to good disciplinary and social impact questions. But getting to such questions requires not "know-it-allness" but constant project listening and open-mindedness. And this includes rethinking your intuitions! Along the way, the research designers we have worked with, including Haven, needed to iterate and revise their originally intuited relationships, but in doing so, and engaging in intentional exercises and reflections, they discovered project scope, significance, and the question that tied them together. In the end, this question may be surprising.

So, take a moment to meditate on staying curious and orienting to the unknown with wonder. Remember that to ask a question is to not know but to be excited about the possibilities that open when you simply ask! This doesn't have to be a process of anxiety or confusion; it can be an experience of awe and anticipation. Orienting in this way is often lost to us in hegemonic or capitalized institutional settings, where questioning and wondering can become interrogating and extracting for utilitarian reasons.

We recommend the following meditation to get back in touch with curiosity and openness. This will help you notice your reactivity and foster interest and excitement for posing an ovearching "don't know" question. Read the steps first and then see if you can try it out.

1. *Find a quiet place to sit.* Allow yourself to be comfortable and alert. Tune in to the external sounds and feel of the space you're in. Feel your contact with your chair or floor. Spend a few minutes noticing the rise and fall of the breath in the chest or belly area.

2. *Remember the last time you felt really curious.* This can be now or in the past. For example, a time when you read a new book, went to a new place, learned a new skill, saw a new film, tried a new recipe, or studied a photo or picture of something you know nothing about, a new form of life or process in the world that you wanted to learn more about. Remember that moment? Feel into the quality of curiosity and wonder and not-knowing.

3. *Sit with that for a few minutes.*

4. *Finish and reflect.* When you're done, gently attune back to your environment. Write down any thoughts, perceptions, and insights about curiosity and the unknown.

Here is how this meditation can help: as when you were curious in the past, you are *feeling into the quality* of a research question, into the wonder of not-knowing, of learning new things together with people and other beings, of exploring potentials and possibilities. Take a moment to think about this. To ease yourself into this process, to let go of conclusions and be open to questions, consider the range of interrogative terms that you might mobilize to create a question from.

This meditation is designed to help you inhabit a creative space of not-knowing. If you want to do a more formal exercise, we recommend reviewing Joe Dumit's Implosion Project, which is an exercise on examining your project in all its aspects.[1] In particular, after answering all the questions, we recommend his wonderful "Twist 2: Gap Map → Ignorance Map." This exercise requires you to make explicit what you do not know about your project. This is extremely helpful for research designers who are either new to their projects or already deeply embedded in their fields.

EXERCISE 7C Draft a Scoping Question

Scoping questions are big questions that bear on scholarly and social transformations. Their answers have ramifications beyond your project. Therefore, answering scoping questions are vital to anthropology and social worlds. Their answers provide ways to understand what "anthropos/human" is as a subject of knowledge-making and as a being-in-relation and a being-in-process. When you write your research proposal, your scoping question will sit at the beginning of the narrative in the first or second paragraphs. If necessary, you can also derive a hypothesis from a good scoping question. It, is, overall, the question that aligns your reader with your whole framework of inquiry.

To create a powerful scoping question, we offer a formula. An effective multidimensional scoping question always has these characteristics:

- Includes one or more MO terms
- Is shaped by several broad, specific scoping terms

Therefore, we represent the scoping question in this formula:

Scoping Question = MO + Scoping Terms

The order of this formula is important to consider when first constructing your scoping question. The easiest part is starting with your MO's terms as an anchor.

Remember that the MO contains both empirical and theoretical concepts and reflects the multidimensionality of your project. The scoping terms also reflect multidimensionality in that they are the broadest but also most specific dimensional aspects of your project. Putting the MO in relation to your scoping terms allows you to see how the empirical scope of your project is connected to a broader context and theoretical generalities. The question that emerges using these concepts does not only represent a research focus; it provides a provocative overarch for fieldwork with potential significance. After you can see this invocation and pattern, the wording and order of terms do not have to be formulaic! After you get the hang of constructing a scoping question, the order of these parts can be tweaked.

The best way to show you the relationship between the formula parts is to gather examples together. Example 7.2 demonstrates the dynamic congruence between MOs, scoping terms, and potential scoping questions.

EXAMPLE 7.2 Scoping Questions = MO + Scoping Terms

MO	Scoping Terms	Scoping Question
Kimberley D. McKinson Security ecologies	Home security; Kingston, Jamaica; Black embodiment; postcolonial and slavery-era artifacts; technologies and practices	How do Jamaican home security practices illuminate experiences of contemporary Black embodiment and legacies of imperial control?
Forest Haven Food sensing	Alaskan settler state, Native Alaskans, regulators, food, sensing, settler colonialism	How do conflicts over Alaska Indigenous food as sensory material reveal continuities and changes in the ongoing project of settler colonialism?

Jason Palmer Im/mobile memberships	Peruvian Mormons, religious pilgrimage, Mormon cosmos, citizenship, US border politics, soul	How do Peruvian Mormons' experiences of soul inclusion in a Mormon cosmos but physical exclusions from Zion (Utah, US) reveal the contradictory dimensions of transnational social memberships?

When it emerges, the scoping question is elegantly simple but intriguingly complex. And it's not easy to land on. But, like the MO, you will feel the rightness of its generative arc in your body.

To illustrate our point, let's take a moment to understand the difference between a well-scoped question and an underscoped question. In other words, when is a scoping question *not* demonstrating broad but specific multidimensionality and tensegrity *and* is simply a more narrowly focused "data-gathering" question (as discussed in this module's introduction)?

Examples 7.3 through 7.5 provide comparisons of strong scoping questions with underscoped or data-focused questions. Taken together, the

EXAMPLE 7.3 Well-Scoped vs. Underscoped Questions: McKinson

MO	Well-Scoped Question	Underscoped Question	What Is the Difference?
Security ecologies	How do Jamaican home security practices illuminate experiences of contemporary Black embodiment and legacies of imperial control?	How do Jamaican security systems help illuminate the politics and discourses of criminality in Kingston, Jamaica?	*The well-scoped question* home security as a materialized practice (*MO term*) to national security policies, slavery, carcerality, differential spatial experiences of Black embodiment and mobility. *The underscoped*

| | | | question gestures toward data gathering because it turns security (MO term) into an abstract process that is only connected to criminality—a term not associated with the project's scope. However, field-based research needs to account for the relationship between security and criminality, making the underscoped question a tool for a data-gathering question. |

EXAMPLE 7.4 Well-Scoped vs. Underscoped Questions: Haven

MO	Well-Scoped Question	Underscoped Question	What Is the Difference?
Food sensing	How do conflicts over Alaska Indigenous food as sensory material reveal continuities and changes in the ongoing project of settler colonialism?	How does Alaskan state regulation obstruct or facilitate Native food-gathering practices?	*The well-scoped question interrelates Indigenous food processes, historical and contemporary political processes, and settler colonial places. The underscoped question gestures toward data gathering* because it asks how regulatory practices directly impact Native

			food gathering. This is a field-based question that will generate data about food gathering specifically. It does not constitute the scoping aims of the project.

EXAMPLE 7.5 Well-Scoped vs. Underscoped Questions: Palmer

MO	Well-Scoped Question	Underscoped Question	What Is the Difference?
Im/mobile memberships	How do Peruvian Mormons' experiences of soul inclusion in a Mormon cosmos but physical exclusions from Zion (Utah, US) reveal the contradictory dimensions of transnational social memberships?	How do Peruvian Mormons negotiate their desires and often inability to migrate (and make pilgrimage) from Peru, one sacred place, to Utah, Mormonism's Zion?	*The well-scoped question* connects the membership (*MO term*) of bodies and souls to sacred places, national places, social and political identities, and worlds. *The underscoped question gestures toward data gathering* because it wants to get information about people's religious motivations to move across borders. As a field-based question, it does not attend to the project's overall scope pertaining to how the soul is hailed via religious belonging and US border politics.

examples show that a strong scoping question is the gateway to the project's overall empirical and theoretical purpose, as well as providing the overarching inquiry structure for the connecting and interacting zones you will engage next.

Because the scoping question isn't intended to activate a particular data-collection process, it must be overarchingly intentional in other ways. As we stated in the introduction to this module, the scoping question must represent your intention to make a disciplinary contribution as well as to elaborate the dimensional possibilities that happen when ethnography creatively connects theory, structure, and events. It must also show intention to link the scholarly and ethical motivations of your project. This requires you to reflect on what an overarching research question can and should do in the world. Asking what it is that we want to know and why is relationally ethical and political. This is why, as we show you in the exercises that follow this one, there is a relationship between the scoping question and your project's significance statements.

Now it's time to draft a question using your scoping terms.

1. *Choose active inquiry words*: As you can see in the examples, big research questions use opening phrases that indicate how the project is asking about a shift in space, time, experience, or another kind of social orienting category. For example, many questions start with "This project asks how" or "This project examines." They then use verbs that signal relationships between continuance and transformation and that connect the forces that compel or alter those relationships, such as re/make, shape, relate, change, reconfigure, shift, structure, impact, reveal, transform, and rethink, among others. If you need further inspiration, you can return to your favorite ethnographies that you selected in Module 2 and look for their central questions and attend to the composition of those queries.

2. *Add a scoping question to Table 7.1*: Using the scoping question formula, scoping terms, and inspiration from Examples 7.2–7.5, draft your question. Don't worry about having the perfect question right now—we provide a reflection process to help you fine-tune it.

3. *Reflect and assess: Is it really a scoping question?* Now ask yourself the following reflective questions about your drafted question:

a. *Can you imagine the answer to this question to start with "yes" or "no?"* Does this question actually inquire about only one or two specific processes that can be answered by doing data collection, such as interviews or archival work?

    i. *Yes.* If the answers to one or both of these questions is affirmative, then it is most likely a question that exclusively guides an aspect of the fieldwork you will do. But it doesn't "overarch" all your fieldwork. As such, it isn't a question that, when answered, will contribute to overarching project scope and significance as we have defined them. If you're having trouble, go back to the Scoping table and keep playing with possibilities. You can also check in with your group members for help.

    ii. *No.* If the answer to both questions is "no," then it is a potential scoping question. Go to b.

b. *Can this question be answered only* after *analyzing data relevant to* all *or most of the peoples, parts, places, processes, contexts, and other disciplinary concepts in the Key Concepts table?* Does it gesture to why answering it might matter to other researchers and to communities at large? Can you imagine it coming at the end of your RD?

    i. *Yes.* If the answers to all these questions are "yes," then it is probably a scoping question or on the way to being one! Because scoping questions are grounded in empirical specificities but pitched at the theoretical or general level, they require theorizing data to answer.

    ii. *No.* If the answers to one or more of these questions is "no," then you may still have a scoping question in process, but it's not broad enough or anchored enough. As a result, it may not be a question that successfully overarches your key concepts. Try to tweak it so that you can say "yes" to the questions above.

**Identify Disciplinary and Social
Significance**

After you compose a scoping question, test it by generating significance statements and reflecting on the relationship between these two project elements. As we indicated in the introduction to this module, a clear statement of project significance communicates disciplinary value but also conveys what would be lost intellectually and socially if your project were never completed. Significance statements are the necessary raw material needed for writing grant proposals, composing dissertation and book proposals, and creating publications and other works. You can get to the significance statements by relating your scoping terms to your literature work.

To understand your project's contributory significance, you must be able to describe your contributions to two major value categories: disciplinary and social. The first reflects your contribution to conversations in the literature. Not only because there is a topical gap (which is significant but not as significant as it could be!) but also because you are multidimensioning project concepts in a way that is novel and needed in those literature conversations. The second aspect of significance, social significance, is how you imagine your work contributing to social transformations. These include, but are not limited to, surfacing the lived experiences of people you are working with, catalyzing political changes, and supporting the emergence of otherwise ways of being and relating.

Before you begin this process, refer to Example 7.6 to see how significance emerged in Forest Haven's project. With an eye for using scoping terms

EXAMPLE 7.6 Haven's Scoping Table

MO + RD	*Food sensing*
	Given ongoing conflicts between state authorities and Alaskan Native peoples, this project examines how twentieth-century assimilation policies still inform the state regulation of Indigenous food practices. It focuses specifically on the sensory dimensions of Native food gathering, hunting, and processing. It positions sensing as a site through which forms of knowing and sovereignty are negotiated and

	enacted. In this way, the project can examine food sensing at the level of everyday embodied Native food practices and the extreme technical practices of Alaskan state authorities regulating Native plants, animals, peoples, and practices. This project aims to analyze the persistent strategies of Native people to respond to shifts in the ongoing project of settler colonialism.
Key Concepts	4Ps: *Persons*: Alaskan Native Peoples, Alaskan state regulators and authorities *Parts*: gathered foods, hunted foods, licenses *Processes*: Native subsistence food practices, state regulation of subsistence practices, food procurement and processing, sensing *Places*: Alaska, Native Alaskan communities, state offices Broad Context: sensory regulation and control, historic and contemporary Indigenous-state tensions, ongoing assimilation efforts of the Alaskan settler colonial state, sovereignty Other Key Concepts: embodiment, settler colonialism, indigeneity
Scoping Terms	Alaskan settler state, Native Alaskans, regulators, food, sensing, settler colonialism
Scoping Question	How do conflicts over Alaska Indigenous food as sensory material reveal continuities and changes in the ongoing project of settler colonialism?

to help determine disciplinary and social significance, review her scoping terms in the table. Take a moment to reflect on how Haven's multidimensional work, focused on "sensing," is both intellectually and socially valuable. Notice how the scoping terms convey the disciplinarily salient importance of studying ongoing but changing food-regulation and resistance processes in Alaska. These processes take place in the context of the Alaskan state's technological changes and climate changes, but they are also broad enough to be relevant to scholars and activists working at junctures of colonial and Indigenous or otherwise practices.

Given these valuable contributions, Haven can consolidate her project's significance in the form of two clear "statements" about its literature-based intellectual and social significance. A table aimed at assessing Haven's project scope and significance might look like Example 7.7.

EXAMPLE 7.7 Haven's Scope and Significance

MO + RD	*Food sensing* Given ongoing conflicts between state authorities and Alaskan Native peoples, this project examines how twentieth-century assimilation policies still inform the state regulation of Indigenous food practices. It focuses specifically on the sensory dimensions of Native food gathering, hunting, and processing. It positions sensing as a site through which forms of knowing and sovereignty are negotiated and enacted. In this way, the project can examine food sensing at the level of everyday embodied Native food practices and the extreme technical practices of Alaskan state authorities regulating Native plants, animals, peoples, and practices. This project aims to analyze the persistent strategies of Native people to respond to shifts in the ongoing project of settler colonialism.
Scoping Terms	Alaskan settler state, Native Alaskans, regulators, food, sensing, settler colonialism
Scoping Question	How do conflicts over Alaska Indigenous food as sensory material reveal continuities and changes in the ongoing project of settler colonialism?
Significance Statements: Disciplinary and Social	1. This project is uniquely designed to research food as a sensed material rather than as an ingested metabolic material. When complete, this project will highlight how a contemporary politics of food sensing is emerging as Native peoples and the state engage in technologically mediated conflict over the regulation of Indigenous food

procurement in a changing environment. This interrelates studies of Alaska, food, politics, science and technology, Indigenous worlds, sensing, and the state in order to broaden an understanding of the sensorial dimensions of Alaskan settler colonialism and Indigenous embodied and spatial sovereignty.

2. This study will contribute to and inform social understandings of the impacts of spatial and embodied food regulation and the Indigenous forms of embodied and spatial resistance.

Note how the scoping question also gestures to the significance of answering it: it uses a focus on the senses to put literatures into a new dialogue based on the need, as Haven identified in her literature searches, for more attention to food as something other than a metabolically political material. The project will therefore inquire about how Native peoples experience the state's sensory mechanisms and how they find ways to maintain their sensory sovereignty. In terms of *social significance*, the question illuminates areas in which social groups are enacting leverage against forms of technical systemic racism and land-based control being unequally deployed under the guise of environmental resource management and climate change adaptation. In sum, answering this question opens otherwise possibilities enacting forms of embodied refusal and sovereignty.

To get yourself to the place of being able to articulate both kinds of significance statements, return to the final literatures you analyzed in Modules 2 and 4. Look back at the review articles and other texts you consulted. Answer the following questions. Notice that some questions repeat a previous exercise—this is intentional and meant for you to notice what has changed or evolved when asked to answer them again. You'll need to use more words to fill out the blanks in this exercise than you will when you draft more pithy significance statements. That's okay! For now we're simply helping you identify why your project is multidimensionally unique and worthwhile in relation to what's come before.

First, develop your disciplinary significance
by following these prompts:

1. *Write out short answers to these questions as succinctly as you can.*

    a. What *established* questions or problems in this literature does your study address?

    b. What *emergent* or as yet unaddressed questions or problems in this literature does your study address?

2. *Fill in the blanks of this disciplinary significance statement guide.* This can be difficult and may require consultation with your advisers or design colleagues, but this is where you can establish how much you can express at this point. Pay attention to whether it is easier to fill in the first blank rather than the second. You may also draft your own significance statement clauses as you see fit.

    a. *Established problems*

        i. This study addresses/interrelates the *established* anthropological topics/problems of _____. This is important because [explain what is uniquely multidimensional about your approach to this established problem] _____ rather than _____ [summarize why this conceptual connection differs from established approaches to this topic/problem].

        ii. By doing work that interrelates/rethinks the relationship between [literatures] _____ it will be important to anthropology because [why linking these literatures is important] _____.

    b. *Emergent problems*

        i. This study addresses/interrelates the *emergent* anthropological topics/problems of _____. This is important because [explain what is uniquely multidimensional about your approach to this emergent problem] _____ rather than _____

> [summarize why this conceptual connection is innovative in light of this emergent issue].

    ii.   By doing work that interrelates/rethinks the relationship between [literatures] _____ it will be important to anthropology because [why linking these literatures is important] _____.

If you find it easiest to fill in statement 2a than statement 2b, be aware that your project may be a *case study*. A case study consciously builds and elaborates existing theoretical work without much new theoretical intervention. You either need to make a stronger argument as to why this particular case study is needed or do further literature review work to discern what makes your study of an established process unique and necessary. Now explore the other dimension of significance: the social.

Now develop your broader social significance
by following these prompts:

1.   *Why is your project important to do now?* Look through your work. If you are having trouble, you can return to these questions—again, allow project listening to guide you:

    a.   Why does this project matter to you?

    b.   Why might it matter to your interlocutors?

    c.   In general, why does this project matter socially and politically, and in what broad and specific ways?

2.   Fill in the blanks of this broader social *significance statement guide*. However, you can draft your own statement clauses as you see fit.

    a.   This study will contribute to understandings of the [experience/process/problem of your research focus] _____ by [justify how and why it will uniquely contribute and to whom it might be valuable] _____.

3.   *Convert the statements above into two cogent significance statements.* Using your statements above, craft a two-part significance statement and enter it into Table 7.2. Don't worry if this

TABLE 7.2 Scope + Significance

MO + RD	
Scoping Terms	
Scoping Question	
Significance State-ments: Disciplinary and Social	i. ii.

is difficult: we follow this work with an exercise to help you troubleshoot the process.

EXERCISE 7E **Confirm the Scoping Question and Significance Relationship**

As you probably found, specifying the significance of your project is challenging. In addition, the project's scoping question and significance have to be clear. This exercise helps you troubleshoot your significance if it is still elusive, then gives you a way to validate the relationship between your scoping question and the project's significance (that is, its overarching scholarly and social value).

So, what should you do if you have trouble with all or any of the significance statements? You may still only have an inkling of significance after doing these exercises—with more concept work and indeed field research it will become clear. However, for the purposes of crafting a compelling scoping question and moving ahead in the exercises, you need to draft some serviceable, if only placeholder, significance statements.

To help with this, do the following question/answer exercises. If you find it difficult to do these alone, turn the questions into a conversation with a person who is capable of recognizing the stakes of the project. This could be an adviser, colleague, or even an interlocutor.

1. *Ask the five "whys."* Buddhist thinker and business leader Sakichi Toyoda invented a way of getting to the root of a problem by asking and answering "why" five times. It helps to be in a quiet and still mode, listening for what feels right. Try a version of this:

   Ask: "Why does this project matter?" Answer "This project matters because…." To each answer, ask again, "Why does *that* matter?" Try to do this for five rounds. You'll notice that each answer will scope out to broader stakes of the project, which is where you can locate significance. Then try to write your significance statements.

If that doesn't work to illuminate significance, try this more personal approach:

2. *Find your personal motivation.* Ask yourself or discuss with a friend or community member, "What are the most compelling aspects of this project to me and/or my community? Why is it important to the community with whom I am conducting research?" When you have answers, continue to ask, "Why is that important?" Keep asking "why?" until you come up to the edge of an a-ha! insight. This may take time and more repetition. Allow room for other questions to emerge beyond "Why is that important?"

After you go through these processes, look together at your Scoping table and see where there is conceptual consonance between your "why" or "personal" answers. This may help you really begin to realize why the project is important in the worlds that it is going to be a part of. If necessary, keep asking yourself questions that continue to dig down to the project's importance.

When you have worked through these prompts, see if you can go back to exercise 7d and generate the significance statements you need. When you get to them, add them to Table 7.2.

3.   *Consider the scoping question–significance relationship.* In Table 7.2, compare your scoping question to your significance statement. Does each significance statement point to the scoping question and reflect the significance of the project? You can check by filling in the blanks of the following prompt, then economizing and smoothing terms. See Example 7.8 following this prompt for inspiration.

The question [insert scoping question] _____ is important because it draws together the broad topics of [insert literatures]. This supports the value of fieldwork to [insert scholarly significance] _____. Researching these processes will inform pressing social and political issues because it will _____.

EXAMPLE 7.8 Haven's Scoping Question–Significance Relationship

The question about conflicts over Alaska Indigenous food as sensory material is important because it draws together the broad topics of Alaska studies, food studies, and science and technology studies. This supports the value of fieldwork to highlight the contemporary politics of food sensing as Native peoples and the state engage in technologically mediated conflict over Indigenous food regulation in a changing environment. Researching these processes will inform social understandings of the impacts of settler colonial food regulation on Indigenous experiences of embodied and spatial sovereignty.

If you cannot complete a similar paragraph (which is necessary in a proposal), revisit your scoping and significance concept work to see where the hangup has occurred. You may also need to bring your half-baked attempt into the group work setting to obtain other perspectives.

EXERCISE 7F Research Project Grid 7

In your Research Project Grid table, add your scoping question and significance statement. For the significance statement, we recommend truncating the two significance prompts used above to create a one- or two-sentence statement that sums up the most outstanding contribution of the project. Then check for consonance across all categories in the project grid.

Collective Concept Workspace 7

You probably experienced in your body how getting into the scoping zone takes you well beyond your RD and into the depth of significance. In other words, you may have felt how your project will make a difference in a variety of ways. While your overall sense of scope and significance *will* change with fieldwork and other research, you have successfully anchored the project into its broader context and its potential to become lively for community and disciplinary importance.

As you do group work, remember that this is just your first foray into scaffolding the inquiry part of your framework. It's an activity that requires mental, corporeal, and analytical flexibility so that as things change or something new is learned, the scoping question itself can easily adjust to those changes. Therefore, consider taking deep, intentional, conscious breaths as you do individual and group work. Consider reviewing your agreements before beginning.

1. In your group, talk through the project design zones. What's clear? Is there any confusion?

2. Share and explain your final Scope and Significance table with your group. Describe the following and solicit feedback from your group:

   a. *Your MO*: Review it again. How does it feel at this point? Does it need adjustment?

   b. *Your scoping terms*: Do they seem congruent with your MO and concept map?

   c. *Your significance statements*: How do they feel? How well do they work?

   d. *Your scoping question*: Does it feel like it captures (part of) the MO and that it turns the concept map into a framework of inquiry?

e.   What relationship do you detect between the scoping question and your significance statements?

f.   How does each concept-work mode feel in the body?

3.   If you are feeling stuck, focus on that and get feedback from the group. Help each other to figure out what needs attention. Talking through the stuckness in a group can accelerate getting clarity on your project.

   a.   If it appears that your MO needs adjustment, and that this is why you can't land on a scoping question, then work together on that. You can draw on your research imaginary, RD, and concept map. You can also redo the MO exercises. These and other exercises can help you to see the entirety of your project and the multidimensional openings that are possible.

   b.   If your overall scope, as represented by the scoping terms, is unclear, then compare those terms with what is in your concept map. Is there anything still confusing, unclear, or still in need of clarification on the concept map? If so, work through those elements and then return to the scoping question again.

   c.   If the significance needs to be further addressed, then talk through the importance of the project to you, to potential interlocutors, and to anthropology. Consider returning to the significance exercises together. These work especially well with at least one other person asking the questions about project importance and meanings.

   d.   If your scoping question isn't landing well, then go through the scoping-question reflection as a group. Together you can also look closely at your Scope and Significance table, which may reveal new insights. Especially check to make sure that at least one part of your MO and the scope of the project are articulated together.

The Connecting Zone

With your scoping question developed, you can now address the process of data gathering. While understandings of "data" and "gathering" are particular to ethnographers and projects, all field-based research aims to learn about and document people's experiences in new and changing spaces. Ethnographic projects need questions that guide those processes of learning and documenting.

As with all aspects of multidimensioning, the process for developing data-gathering questions centers the power and potential of concept *connectivity*. Although your disparate project concepts are distinct from each other, they can cluster into broad processes that you can inquire about. Understanding those broad *process clusters* will help you to articulate social scientific, as well as otherwise, justifications for why your inquiry is important. You can then develop corresponding *data-gathering questions* to guide that field research. Given these design parameters, you can see why we distinguish the scoping question as theory focused and the data-gathering questions as empirically focused. These empirically focused questions will orient your work and life in the field and connect your overarching theory-focused question to your project's fieldwork plan, broader significance, and otherwise intentions. The connective framework of this inquiry zone, then, plays a crucial role by holding the ideational and practical dimensions of your project together, which is why we call it the *connecting zone*.

MODULE PURPOSE
Form process clusters and questions that will provide the data needed to answer and theorize the scoping question.

An evocative way to think about this zone is to imagine it as bodily connective tissue that fleshes out inquiry-based tensegrity. The connecting zone conjoins your project's distinctive parts, materializes its major functions, and allows you to enact your curiosity and intentions. The concept clusters and data-gathering questions you will create through this process advance the overall aims of the project by connecting socially intimate and extensive activities with broader contexts.

This zone is also where the MO, RD, project elements, and significance get connected to the on-the-ground process of fieldwork. We provide examples throughout this module that show you how to iterate these connective project features if necessary and keep them congruently interrelated. As with all modules, this concept work might help you refine your previous work. For example, the act of clustering concepts into processes that you will investigate can illuminate whether your scoping question and significance accurately point to what you're going to do in the field. Iteration is your friend!

Here is a summary of this zone's activities. In the first exercise, you'll use your concept map to identify three to five conceptual *process clusters*. Although you already made connections among individual concepts when you mapped your concepts and their connections (Module 3) and developed your multidimensional concept combos (Module 4), this new exercise allows you to cluster key processes that become avenues for data collection. You can then define *data sets*: the specific array of information, from archival data to interview data, that you need to answer your scoping question. This will help you articulate data-gathering strategies, which are necessary for proposal writing. In the second exercise, you will develop three to five data-gathering questions that correspond to these process clusters and data sets. The data-gathering questions do two things: (1) they define and justify your claims about where you need to go and what you need to do there; and (2) they make visible the kinds of ethnographic material you need, at the end of your project, to write, theorize, and articulate its value. All of this is necessary to define a robust and dynamic multidimensional proposal activity plan.

Exercises

EXERCISE 8A Identify Your Process Clusters

The three to five process clusters that you identify will channel your framework of inquiry into productive fieldwork data-collection avenues. Remember: the purpose of clustering these concepts is not only to create those data-collection avenues, but to show how the innovative multidimensionality of your project will be put into action. The clusters bring your 4P, contextual, and other key concepts into the kinds of relationships that enact your innovative aims. This may require several rounds of playing with clustering and reclustering. You may end up amending or reaggregating those clusters as you consider whether they will support the multidimensioning you articulated in your RD, confirmed with your concept combos, and which is invoked by your MO.

The next step is to put these process clusters into a table and name them. How you name these clusters is very important for multidimensional work because they remind you of what is being connected and why. As you do your process clustering, you need to stay mindful of your MO and scoping question. Therefore, you want to include them in a Process Clusters table (see Table 8.1). Does this seem challenging? It is, but when you achieve it, you will clearly see what it is you will be doing when you do fieldwork.

Before you begin, look at Examples 8.1 and 8.2. They show how the clusters create a sense of a project's main conceptual dimensions. They interlock people + processes + sites + contexts + other concepts in ways that, when grouped and viewed together, illuminate the overarching scope and connective novelty of the project. Notice how some concepts are shared between process clusters.

EXAMPLE 8.1 Wilkinson's Process Clusters

MO: Securing the family

RD: The movement against "gender ideology"—a concept used by right-wing groups globally to reference the social construction of gender—is gaining force across Europe and Latin America. Over the past decade, pro-family movements have claimed that gender is rooted in nature and that this scientific fact must be defended. This allows right-wing organizations in Mexico to claim that they must secure the family against gender equality and LGBTQ rights. In doing so, the Mexican pro-family movement makes "gender ideology" a national security problem that bears on state corruption, violence and crime, and neocolonialism. Using ethnographic research, the project follows how right-wing theories of the family connect to theories of the state. It also investigates how gender and security agendas are being shared among authoritarian populist movements across the globe.

Scoping Question: How does the Mexican pro-family movement's involvement in gender ideology activism, emerging in the context of political and economic security crises, illuminate contemporary forms of transnational right-wing organization and theory?

Key Concepts on Map	Process Cluster Name
– Gender – Gender ideology – Conservative family – Dehomosexualization – LGBT activists – Conservative science – Antiscience – Security threats – Corruption – Government equality acts	Gender
– Security discourses – Mexican state securitizations – Insecurities – Sovereignty – Equality	Security

– Authoritarianism – The State – Governance crises – Violence – Corruption	
– Right-wing activists – Antigender activism – Pro-family culture – Transnational conservative organizations – Catholicism/Pentecostalism – Traditionalism – Digital political platforms	The Right-wing

EXAMPLE 8.2 Kladky's Process Clusters

MO: Financial salvation
RD: Within deindustrializing Appalachia, there is a growing movement to manage family life using religious financial practices. This project examines intertwined changes to financial subjectivity, family life, and social class through an ethnography of Christian financial saving and literacy programs in the region. Based in working-class communities in economic transition, it follows what happens to white identity and family processes as communities lose historical forms of livelihood, take on debt, and reorient to racial and class privilege. Situated in broader interconnected opioid, environmental, foster care, and job safety crises, the project examines how novel economic practices are changing the concept and experience of class in the United States.
Scoping Question: How do Christian family finance movements in Appalachia reveal the conjoined structural and ideological processes that are transforming US working-class subjectivities?

Key Concepts on Map	Process Cluster Name
- Family identities - Intimacy - Race - Religion - Family finance - Kinship - Foster care - Households - Appalachian family culture - Racialized identities	Family
- Financial savings and literacy programs - Class subjectivities - Debt - Evangelical economic theory - Responsibility - Social (im)mobility - Racialized identities - US class formation	Financial class
- Coal industry - Changes in labor - Unemployment - Deindustrial environments - Postindustrialization policies - Regional economic crises - Social services offices	Deindustrialization
- Regional inequalities - Regional identities - Dirty energy - Rural labor precarity - Labor organizing - Rural poverty theories - White poverty theories - Environmental crises - Opioid crisis	Saving Appalachia

Here is a step-by-step guide to identifying your process clusters and putting them into your table. Read through the steps first to orient yourself to the goals of the exercise:

1. *Update your concept map as needed.* Add and arrange any new concepts developed from both the Key Concepts table and the Scoping table. Appreciate how the concept map arranges and connects the project's major elements in the form of terms and phrases, with your MO, the emblematic connective concept, holding it all together.

2. *Prepare to draw three to five concept clusters on your map.* There are two ways to do this cluster technique with your concept map, depending on what kind of map you have created. One way, with a hand-drawn or printed-out map, is to draw circles around concepts (note that concepts for one cluster may not all be located in the same part of the map). This may lead you to draw circles that overlap, because terms and phrases might contribute to one or more data-collection clusters. This may lead you to redraw your map to facilitate clustering. Or, if you have a digital mapping program, you may move terms and phrases around (sometimes duplicating them) to create the large process clusters. This may require redoing the map in order to facilitate the digital clustering process.

3. *Using the examples as guides, draw your three to five process clusters.* Look for concepts that can be brought together to map large social process categories that focus your data-collection avenues. What are the key concepts that hang together-as a data collection focus—in ways *that your project needs to enact* in order to address the scoping question? Because these are data-collection clusters, meaning that you will engage with people doing things in places, each cluster should include these kinds of concepts (refer to your Key Concepts table):

   a. At least one (but ideally more) of the 4P elements: people, places, parts, processes
   b. At least one big-picture context concept
   c. At least one "other" key concept

4. *Transfer your drawn clusters to Table 8.1 by placing each group of concepts into a "Key Concepts on Map" box.* When complete, you should have three to five boxes filled out.

5. *Name your clusters.* Think about distinct names that reflect how and why your concepts are grouped and the major processes they signify. Try to use one word to name the cluster, but if you need two or three that is fine. You want to be able to see, at a glance, the appropriately bounded but broad processual range of your fieldwork world.

Now audit your work for connections to your MO, RD, and scoping question:

6. MO *connections*: Do the clusters, and their names, seem to be key aspects of your MO? If you imagined that you could look your MO up in a dictionary, would these clusters, or their names, be listed as examples of how to understand what the MO is?

7. RD *connections*: Most if not all the clusters should be perceivable as *processes*, or combinations of processes, that your RD points to. If not, ask yourself whether the RD and MO need to be revised. If so, take time to redo exercises in Modules 5 or 6 as needed to tweak them. This could be a moment when you see a new dimension to your project that could be reflected in your MO, RD, or even your literatures.

8. *Scoping-question connections*: Imagine the data that you are going to get from your clusters, which typically come in the form of interactional or interview data, observations, or archival research. Will the data in these clusters be enough to answer (and ultimately theorize) your scoping question? If so, you have a good set of clusters. If not, consider what is missing and if you need to amend a cluster or two or reaggregate concepts into a new grouping.

TABLE 8.1 Process Clusters

MO:	
RD:	
Scoping Question:	
Key Concepts on Map	**Process Cluster Name**

Data sets are a definitive list of everything that needs to be collected to answer a data-gathering question (which you will develop in Exercise 8c).

So, why don't you identify data-gathering questions *first*, and then describe the data sets? The reason is that we want you to stay true to your concept map, which you've validated several times so far, as the best source of
an overall view of processes and elements that you need to find out about.
Researchers can identify data-gathering questions without the benefit of this
overall picture of interrelated key concepts. This means they might foreclose
understandings of what the key process clusters are, as well as the data that
is needed to illuminate them. As you work further in the module, you'll find
yourself tacking back and forth between the process cluster, data set, and
data-gathering question formulation process. For now, we're focusing on
bringing your data-gathering needs into view. Look at Examples 8.3 and 8.4
before you go any further.

EXAMPLE 8.3 Wilkinson's Process Clusters and Data Sets

MO: Securing the family
RD: The movement against "gender ideology"—a concept used by right-wing groups glob-ally to reference the social construction of gender—is gaining force across Europe and Latin America. Over the past decade, pro-family movements have claimed that gender is rooted in nature and that this scientific fact must be defended. This allows right-wing organizations in Mexico to claim that they must secure the family against gender equality and LGBTQ rights. In doing so, the Mexican pro-family movement makes "gender ideology" a national secu-rity problem that bears on state corruption, violence and crime, and neocolonialism. Using ethnographic research, the project follows how right-wing theories of the family connect to theories of the state. It also investigates how gender and security agendas are being shared among authoritarian populist movements across the globe.
Scoping Question: How does the Mexican pro-family movement's involvement in gender ideology activism, emerging in the context of political and economic security crises, illumi-nate contemporary forms of transnational right-wing organization and theory?

Key Concepts on Map	Process Cluster Name	Data Sets
- Gender - Gender ideology - Conservative family - Dehomosexu-alization - LGBT activists - Conservative science - Antiscience - Security threats - Corruption - Government equality acts	Gender	- The genealogy and right-wing the-orization of "gender ideology" as a social, political, and anti-Christian construction - Activists' analysis of gender as a secu-rity threat, as neocolonial imposition, and as linked to state corruption - State and legislative discourses on equality, women's rights, and LGBT communities
- Security discourses - Mexican state securitizations - Insecurities - Sovereignty - Equality - Authoritarian-ism - The State - Governance crises - Violence - Corruption	Security	- Broad activist discourses and practices that link gender and threats to family and state security - How activists personally connect indi-vidual, family, and state security - How the right-wing theorizes Mexican national crises - Mexican state security policies in ev-eryday (education, labor, private property) and national processes (eco-nomic, military, migration) - Histories of the Mexican state and re-ligious understandings of relationship between family and sovereignty
- Right-wing activists - Antigender activism	The Right-wing	- How people become right-wing activists - How activists act to connect church and other key institutions and organizations

– Pro-family culture – Transnational conservative organizations – Catholicism/ Pentecostalism – Traditionalism – Digital political platforms		– The relationship between Mexican pro-family organizations and transnational pro-family organizations—and how they build and exchange information – Activists' worldviews and political analysis regarding truth, liberalism, science, as well as information-production strategies and logics – How activists build communication platforms (social media, church networks, etc.) to enable mass mobilization

EXAMPLE 8.4 Kladky's Process Clusters and Data Sets

MO: Financial salvation

RD: Within deindustrializing Appalachia, there is a growing movement to manage family life using religious financial practices. This project examines intertwined changes to financial subjectivity, family life, and social class through an ethnography of Christian financial literacy programs in the region. Based in working-class communities in economic transition, it follows what happens to white identity and family processes as communities lose historical forms of livelihood, take on debt, and reorient to racial and class privilege. Situated in broader interconnected opioid, environmental, foster care, and job safety crises, the project examines how novel economic practices are changing the concept and experience of class in the United States.

Scoping Question: How do white Christian family finance movements in Appalachia reveal the conjoined structural and ideological processes that are transforming US working-class subjectivities?

Key Concepts on Map	Process Cluster Name	Data Sets
– Family identities – Intimacy – Race – Religion	Family	– Family social histories and experiences of transitions to postindustrial work modes – How households create a spiritual and salvational relationship with income, saving, and debt

– Family finance – Kinship – Foster care – Households – Appalachian family culture – Racialized identities		– Appalachian social history – Governmental and alternative archives – Racial and gendered logics of intergenerational family life, labor, and the "white family" – Affective and spiritual qualities of nuclear and intergenerational ties to finance and household support
– Financial literacy and savings programs – Class subjectivities – Debt – Evangelical economic theory – Responsibility – Social (im)mobility – Racialized identities – US class formation	Financial class	– How people understand class, race, and the "white worker" – Experiences with personal and regional financialization processes – Community and individual economic histories of income and debt – Christian financial literacy program recruitment, retention, instruction – Discourses of personal and vocational responsibility – Histories and contemporary iterations of racialized working-class identity
– Coal industry – Changes in labor – Unemployment – Deindustrial environments – Postindustrialization policies – Regional economic crises – Social services offices	Deindustrialization	– Histories of industrial social structures and labor movements – Deindustrialization processes and policies – General trends in postindustrial work, employment, wages, and family life – Experiences of economic in/security in relation to changes in work type and availability – Historical and contemporary organized labor and decreasing power of unions – Changes in safety-net infrastructure: benefits, pensions, mortgages

	Saving Appalachia	
– Regional inequalities – Regional identities – Dirty energy – Rural labor precarity – Labor organizing – Rural poverty theories – White poverty theories – Environmental crises – Opioid crisis		– Regional socioeconomic formations and shifts – Community senses and making – New forms of regionally defined labor like short-term contract-based jobs – Regional crises: opioid, environmental, foster care, and occupational safety – Longitudinal experiences and dynamics of racial regionalities

After you have reviewed the examples, try this process yourself and record your findings in Table 8.2. Here's how to identify and build your data-set descriptions:

1. *Feel the need to learn.* For each of the three to five process clusters you identified in the previous exercise, ask yourself these questions: What are all the things I need to learn about to understand this connecting cluster process or processes in order to work toward answering my scoping question? In research language, what are the kinds of data I need to gather?

2. *Feel the connectivity between the MO, RD, scoping question, and data sets.* Describing your data sets is a crucially connective moment for your multidimensional project. To validate that your data set is indeed multidimensionally informed, make sure to hold your MO and scoping question in view. Reflect on whether the data set that you're proposing is necessarily and strongly connected to your theoretical object of study (the MO), your short and sweet description of what you're doing and why (RD), and the question you want to answer when you're done with

TABLE 8.2 Process Clusters and Data Sets

MO:		
RD:		
Scoping Question:		
Key Concepts on Map	**Process Cluster Name**	**Data Sets**

fieldwork and are working with the materials you got (scoping question).

3. *Verify the cluster names.* Check that each cluster name also describes your data sets. If necessary, revise the cluster name to better connect what you're clustering with what you're gathering.

Next, you'll identify the questions that connect the process clusters with the data sets they represent.

Data-gathering questions point to the need for data sets. They also orient and direct your data collection. They specify the fieldwork-based how/who/what/where/when that your scoping question opens up. As you can see now, having done the work above, these questions must be posed in relation to the people and beings you want to interact with. They might be focused on particular parts (things) that matter in your project. They are anchored in key social spaces (a lab, a bureaucratic agency, activism, a religious center, etc.). They usually address a sense of socially demarcated processes of knowing, experience, and relating. The most successful proposals pose their data-gathering questions soon after they specify their most overarching project aims and question.

When aggregated, the data-gathering questions should represent all elements of the concept map and should give a sense of the scope of the project. Now you may see why we do data sets first, and questions second. However, you must still tack back and forth between these exercises, tweaking the elements in the table to make sure your project congruence and tensegrity are strong.

In Examples 8.5 and 8.6, notice that the data sets create multidimensionality out of the process cluster by including ethnographic elements (4Ps) as well as the broader contexts and processes that situate them. The

EXAMPLE 8.5 Wilkinson's Data-Gathering Questions

MO: Securing the family
RD: The movement against "gender ideology"—a concept used by right-wing groups globally to reference the social construction of gender—is gaining force across Europe and Latin America. Over the past decade, pro-family movements have claimed that gender is rooted in nature and that this scientific fact must be defended. This allows right-wing organizations in Mexico to claim that they must secure the family against gender equality and LGBTQ rights. In doing so, the Mexican pro-family movement makes "gender ideology" a national security problem that bears on state corruption, violence and crime, and neocolonialism. Using ethnographic research, the project follows how right-wing theories of the family connect to theories of the state. It also investigates how gender and security agendas are being shared among authoritarian populist movements across the globe.

Scoping Question: How does the Mexican pro-family movement's involvement in gender ideology activism, emerging in the context of political and economic security crises, illuminate contemporary forms of transnational right-wing social organization and theory?

Process Cluster Name	Data Sets	Data-Gathering Questions
Gender	– The genealogy and right-wing theorization of "gender ideology" as a social, political, and anti-Christian construction – Activists' analysis of gender as a security threat, as neocolonial imposition, and as linked to state corruption – State and legislative discourses on equality, women's rights, and LGBT communities	How do pro-family organizations connect their opposition to gender equality and self-determination to processes of moral and scientific nation-building?
Security	– Broad activist discourses and practices that link gender and threats to family and state security – How activists personally connect individual, family, and state security – How the right-wing theorizes Mexican national crises – Mexican state security policies in everyday (education, labor, private property) and national processes (economic, military, migration) – Histories of the Mexican state and religious understandings of relationship between family and sovereignty	How do right-wing activists theorize family securitization as a way to combat corruption, create national sovereignty, and build spiritual and public well-being?

| The Right-wing | – How people become right-wing activists
– How activists act to connect church and other key institutions and organizations
– The relationship between Mexican pro-family organizations and transnational pro-family organizations—and how they build and exchange information
– Activists' worldviews and political analysis regarding truth, liberalism, science, as well as information-production strategies and logics
– How activists build communication platforms (social media, church networks, etc.) to enable mass mobilization. | How are contemporary Mexican right-wing identities, activist programs, and global solidarities being shaped by transnational coordinations to combat gender ideology? |

EXAMPLE 8.6 Kladky's Data-Gathering Questions

MO: Financial salvation
RD: Within deindustrializing Appalachia, there is a growing movement to manage family life using religious financial practices. This project examines intertwined changes to financial subjectivity, family life, and social class through an ethnography of Christian financial literacy and saving programs in Appalachia. Based in working-class communities in economic transition, it follows what happens to white identity and family processes as communities lose historical forms of livelihood, take on debt, and reorient to racial and class privilege. Situated in broader interconnected opioid, environmental, foster care, and job safety crises, the project examines how novel economic practices are changing the concept and experience of class in the United States.

Scoping Question: How do white Christian family finance movements in Appalachia reveal the conjoined social and ideological processes and structural forces that are transforming US working-class subjectivities?

Process Cluster Name	Data Sets	Data-Gathering Questions
Family	– Family social histories and experiences of transitions to postindustrial work modes – How households create a spiritual and salvational relationship with income, saving, and debt – Appalachian social history – Governmental and alternative archives – Racial and gendered logics of intergenerational family life, labor, and the "white family" – Affective and spiritual qualities of nuclear and intergenerational ties to finance and household support	How do white families act to connect financial practices, religion, and family management as they navigate the afterlife of industrialization?
Financial class	– How people understand class, race, the "white worker" – Experiences with personal and regional financialization processes – Community and individual economic histories of income and debt – Christian financial literacy program recruitment, retention, instruction – Discourses of personal and vocational responsibility	How do white families reimagine and enact concepts of class, as households have transitioned from relatively secure livelihoods to living with debt?

	− Histories and contemporary iterations of racialized working-class identity	
Deindustrial-ization	− Histories of industrial social structures and labor movements − Deindustrialization processes and policies − General trends in postindustrial work, employment, wages, and family life − Experiences of economic in/security in relation to changes in work type and availability − Historical and contemporary organized labor and decreasing power of unions − Changes in safety-net infrastructure: benefits, pensions, mortgages	What are the structural aspects of deindustrialization and how do householders materially and affectively navigate them?
Saving Appalachia	− Regional socioeconomic formations and shifts − Community senses and making − New forms of regionally defined labor like short-term contract-based jobs − Regional crises: opioid, environmental, foster care, and occupational safety − Longitudinal experiences and dynamics of racial regionalities	How have industrial, labor, community, and racial histories shaped contemporary concerns about "saving the family" and "saving Appalachia" in the context of regional crises?

data-gathering questions provide direct and specific data-gathering objectives. They are formulated to capture the tensegrity represented in the process clusters and the data sets.

1. *Fill out your Data-Gathering Questions table (see Table 8.3). Pay attention to each process cluster's dimensions and allow the excitement and humbleness of not-knowing into your thoughts and embodied experience. As you compose these questions, keep in mind that they are research-in-action queries, whereas*

TABLE 8.3 Data-Gathering Questions

Process Cluster Name	Data Sets	Data-Gathering Questions

your scoping question was an analytic query. Free-writing can help here. Questions should include not only the ethnographic aspects of the project but also some background, context, and archival elements of the project.

Now that you constructed your questions, you will check to see if that your concept map, MO, RD, scoping question, and data-gathering table elements are congruent.

2. *Connect the map and questions.* When you look at your concept map, do your data-gathering questions seem to cover the field? If not, what is missing or incongruent? Are your questions amiss or is your concept map amiss? Whatever might be missing on the concept map needs to be incorporated into the data-gathering questions. In addition, adjusting your concept map might mean rethinking the data-gathering questions. Adjust the map or questions as needed.

3. *Ensure element congruence.* Review Examples 8.5 and 8.6. Write your MO, RD, and scoping question above your Table 8.3. Make sure that your questions support the scoping question and RD. Adjust your questions if necessary.

EXERCISE 8D **Research Project Grid 8**

From the tables you have completed, transfer your process cluster names, the corresponding data sets and data-gathering questions to the grid. Check whether these new project elements are playing the connective role between your broader descriptive (RD) and bridging (MO) or overarching inquiry (scoping question) elements. Make some notes about your sense of connection using these prompts. Prepare to discuss them with your group.

1. *Perceive connectivity.* How do your connecting-zone elements help to connect the intellectual and aspirational dimensions of your project with the practical fieldwork dimensions?

2. *Perceive disconnectivity.* Make notes of where you see any disconnection between your connective-zone elements. Why do you feel this sense of disconnection? What do you think you need to do to address it?

Collective Concept Workspace 8

This module was a lot of work! It mobilized all the key concepts you have been assembling and iterating up until now. Appreciate the more nuanced connections that you've now made out of your concept map. While you will do a congruence check for all these categories in the next module's project grid, now is the time to project-listen, to feel for how the clustering and development of data sets and data-gathering questions are landing for you. Consider revisiting your agreements before beginning.

In your groups, exchange your concept maps and versions of Table 8.3. You'll be working through the following:

1. Using your grid reflection above, identify any gaps, confusion, or incompleteness in any of the work in this module or previous ones. Sometimes an exercise for the connecting zone can trigger an insight about your MO, your RD, or your scoping question, among others. Talk through any work that might need to be revisited in order to create stronger congruence across all these project elements.

2. Compare your concept map to your process clusters. Do all your cluster concepts either match or connect with other concepts on the concept map? If anything feels like it's missing or disconnected, ask your group members for their thoughts about these incongruities.

3. Compare the process clusters to the data sets and the data-gathering questions. The data-gathering questions both orient you to the specifics of field research and connect that research to the conceptual concepts. Where do you see congruence, and where might there be gaps? If anything is "off," adjust while in the group. The data sets orient you toward all the information and data you need to gather to answer the

data-gathering questions. Does everything look like it's accounted for? Is there anything else you need to do to learn more about your project and ultimately be able to answer the scoping question?

4. Contemplate whether the data-gathering questions, when assembled together, will contribute to answering the scoping question. If the list doesn't feel right just yet, see what help you can get from group work and/or wait to see what adjustments arise later.

MODULE 9

The Interacting Zone

You just completed the most challenging aspects of multidimensional work: drafting the scoping question, stating your project's significance, and crafting data-collection sets and questions. Now you can turn your attention to planning field-level inquiries. This zone is where you work on your project's *interactions*.

In the *interacting zone*, you will move from thinking about categories of data to focusing on direct forms of engagement. We call these "interactivities" in this module, signaling that all actions you do in the field are interrelational and socially interdependent. *Interacting questions* guide your fieldwork involvement in various social processes (including things you will witness and get involved with), enable work with multimodal materials and archives, and guide processes of engaging and conversing with people.

By generating these interactive questions and methods plans, you'll finalize the framework of inquiry across your project's conceptual and practical dimensions. You will specify the relationships between the broad conceptual aims of your project and the approaches needed to make data. The data you collect from interacting questions will directly contribute to answering the data-gathering questions, which, further on, will help you provide an analytic and theoretical answer to the scoping question.

Generate field-based questions that connect directly to methods; transition to the communicating and planning phases of project work.

The interactive zone is thus the space where your curiosity can be woven into everyday engagements. This is where you'll identify the basic activities you need to write about in a research proposal—to show what you'll be observing, joining, and meeting with people about during a short- or long-term project.

After completing this module, you'll have a fully filled-out research project grid that contains all the components you need for grant, book, and thesis proposals, which we address in depth in Module 10. The specific interacting questions—observational, interpersonal, and archival—may or may not show up directly in your proposal writing. But they will definitely be useful in completing statements of purpose, making agreements with community or business members, or filing ethics protocols.

To enter this zone, we offer a set of questions to orient yourself to the contextual, ethical, and political dimensions of what your fieldwork is about: the three *hows*. Keep these simple questions in mind during your work in the interacting zone. Importantly, these questions are just as relevant when you are designing fieldwork as when you are in the field doing it.

TABLE 9.1 The Three *Hows*

1. How did [ ] emerge?
2. How does [ ] work?
3. How and for whom does [ ] matter?

The brackets that are part of these questions indicate a placeholder for any social process or thing: for example, a yearly ritual, a business practice, a political position, an identity, a social category, a gardening technique, a virtual technology, or a planet-scale media network.

These *how* questions not only help you stay grounded and present with your project aims. They also help you develop distinctive but coordinated approaches to your field activities. Why? Because they hold open inquiry about the livingness of interactive space, times, and meaning-making.

The first "how" is the historical or genealogical question that reminds you to look for the many perspectives on how the processes you're studying came into being. The second "how" is the ethnographic question that keeps you focused on the many working parts of those processes, their conditions, and the ways they operate differently for different people. The third "how" is the ethical and political question that points you toward the personal and social significance of these processes: for groups of interlocutors, for you as the researcher, and for anthropology as a practice and discipline. By keeping these three *hows* in mind, you can devise specific interacting-zone questions and choose the best methods for the specific social processes you plan to engage.

This module is also where you will begin to specify your project's methods. We come to methods last in this handbook because they are meant to be in service to the project conceptualization process, rather than something that determines it. As we stated in the prelude and introduction, this handbook does not teach field methods or the politics of ethical engagement. But we expect that you have been exposed to ethics issues, basic anthropological methods at large, and ethnographic methods specifically. We also assume that you have ways to investigate new and nonstandard methods that can help multidimension your project.

With all this in mind, the archival, observational, and conversational questions you develop, and the ways you'll go about asking and answering them, can be well tailored to your conceptual aims. For example, your activities may involve tried-and-true methods like focus groups. But you might also, in the service of your multidimensional objectives, need to consider new or different approaches to assembling and conversing with groups. Therefore, we included an exercise to help think about *other* methods. At the end of this module, you will reflect on how the questions you formulated in both the interacting and connecting zones are congruent and how that congruence makes the project aims legible.

As you proceed to the exercises in which you draft your interacting questions and select your methods, keep in mind that at any point (even during and after field research) you may need to go back and reengage and reexamine your connecting-zone or scoping-zone elements. For example, you may draft an interacting question for doctors about their practices and then realize you need to ask a lot of questions about physician training; however, medical school processes were not previously conceptualized in your data sets or data-gathering questions. That may lead you to ask whether data sets or questions need to be tweaked, a new data set or new data-gathering question needs to be substituted for a previous one, or new data-gathering questions need to be added.

Overall, the concept work and multidimensional congruence in this interacting zone are just as vital to the project's design as other zones. The questions you pose to interlocutors, to the archive, and when you observe activities are not "lesser" than the broader questions of the preceding modules. Since you, and possibly your project collaborators, will be posing these questions in your everyday interactions, they will intimately help shape your experiences in social worlds. Interacting questions, and the methods you use in the field, manifest the interpersonal dimension and potentials of your project. They provide the means to enact the multidimensional objectives you have explored through all the modules. This is how you end up with a smoothly integrated framework of inquiry.

The exercises begin by helping you identify what interactivities you need to do to answer your data-collection questions. Then you will express those activities in terms of methods (e.g., interviewing, observing) that you need to justify. We call these *interactivities* to help keep the focus on the interpersonal and interdimensional dynamics they create.

Exercises

EXERCISE 9A Establish Your Data Collecting Interactivities

This exercise takes you back to the connecting zone, which you can now access to create guidance for data-collecting interactivities. It's important to make sure you know which data sets require interactivities of interviewing, observing, or archival work. Or, even more importantly, whether they might require other kinds of less common methods. Thinking about interactivities, and precisely how you'll enact them (i.e., methods), requires thinking back to the data sets you need. Example 9.1 will help you see the relationship between your project elements and articulate the kinds of interactivities and methods you will use.

Here are the instructions for filling out Table 9.2.

1. *List data interactivities and methods for each process cluster's datasets.* For each data set associated with your process cluster, list and describe the kinds of interactivities you imagine doing to answer the cluster's data-gathering question. To ensure that you're selecting the right activities, look back at your Key Concepts table. Check to make sure that each of your activities are field-grounded by having 4Ps specifically mentioned: people, parts (things), places, and processes.

2. *Fill out the "why these methods" column.* In the last column of the table, explain how these activities will facilitate gathering the data you need. If you need data for which you don't have a method identified, indicate that as well by describing what you need to know. Example 9.1 represents interactivities for just one data-gathering question and its associated data sets. Notice the

EXAMPLE 9.1 Wilkinson's Data Sets–Data-Gathering Questions–Interactivities Relationship

MO: Securing the family
RD: The movement against "gender ideology"—a concept used by right-wing groups globally to reference the social construction of gender—is gaining force across Europe and Latin America. Over the past decade, pro-family movements have claimed that gender is rooted in nature and that this scientific fact must be defended. This allows right-wing organizations in Mexico to claim that they must secure the family against gender equality and LGBTQ rights. In doing so, the Mexican pro-family movement makes "gender ideology" a national security problem that bears on state corruption, violence and crime, and neocolonialism. Using ethnographic research, the project follows how right-wing theories of the family connect to theories of the state. It also investigates how gender and security agendas are being shared among authoritarian populist movements across the globe.

Scoping Question: How does the Mexican pro-family movement's involvement in gender ideology activism, emerging in the context of political and economic security crises, illuminate contemporary forms of transnational right-wing social organization and theory?

Process Cluster Name: Gender

Data-Gathering Question for this Process Cluster: How do pro-family organizations link antigender ideology discourses and practices to efforts that are intended to make families and the nation "secure?"

Data Sets	Interactivities and Methods	Why These Methods?
Understand the genealogy of "gender ideology" (as social construction, as anti-LGBT) among right-wing groups	Participant observation with the pro-family organizations focusing on the production, analysis, and communication of antigender ideology and its links to concerns about corruption, security, and neocolonial impositions; members of pro-family organizations, especially leaders and longtime activists who strategize antigender analysis as well as hold organizational memory Interviews with pro-family organizational members to establish antigender epistemological genealogies Interviews with pro-family organizational members to establish antigender epistemological genealogies	To get a multidimensional view of how antigender ideology emerges and gets constructed by pro-family organizations; and to learn how and why their analysis includes state corruption, security, neocolonial impositions

Get data about all aspects of activists' analysis of gender as a security threat, as neocolonial imposition, as linked to state corruption	Collect gray literature from pro-family organizations' literature production and analysis (text, video, audio, social media, and their own archives) of gender ideology and deshomosexualización Participant observation at activist meetings and protests	To understand how pro-family organizations produce, analyze, and circulate antigender ideology
Learn about state discourses of equality and legislative efforts to increase (or not) the rights of women and LGBT communities	Archival and historical research on Mexican state responses to gender and LGBT equity	To understand the specific state policies that pro-family organizations are protesting and responding to
Understand how "gender ideology" resonates across other social and institutional spheres	Interviews (and oral history interviews) with pro-family organizational members to understand antigender historical and contemporary connections to Catholic and Pentecostal churches, as well as other aligned organizations and institutions Interviews with pertinent members of those allied institutions Participant observation at allied institutions' meetings or other events pertinent to gender ideology	To understand the history and genealogy of antigender ideology across institutions in order to track the discourse and action of right-wing activism

congruent relationship between the data sets, interactivities and methods, and "why these methods" sections—and how together they help answer and justify the data-gathering question.

Now that you've noted the need to use particular methods, you can draft specific *interacting questions* you will pose to guide your observations, to have conversations with people, and to approach the archives you will work with. Here you need to be very clear and congruent about which data you need, what you are going to see or be involved in, the permissions needed, how you will collect that data, and why it's important. Overall, you need to be able to perceive how these parts connect with each other.

Remember the three *hows*? Effective interacting questions usually revolve around the research commitment to understanding *how*. We begin with observations, in which one is perceiving processes from the position of a nonparticipant or participant. In the following exercises, you will complete

TABLE 9.2 Data Sets–Data-gathering Questions–Methods Relationship

RD:		
MO:		
Scoping Question:		
Process Cluster Name:		
Data-Gathering Question Connected to Process Cluster:		
Data Sets	Interactivities and Methods	Why These Methods?
Add rows as needed . . .		

3 tables. These tables are a record of your design intentions; keep them for the future and compare them to what you actually encounter during field research. Revising your data-gathering and interacting questions is a dynamic process that you can use to help reorient your project midcourse or to help you with multidimensionality.

EXERCISE 9B **Define Questions for Your Observational Interactivities**

You will develop interacting questions for observations—those that you are witnessing and those that you can participate in. Because observations involve your capacity to watch, follow, be a part of, and follow up on, questions to guide them are posed at the level of embodied engagement in a social way: perceiving, joining, being invited, feeling, moving, learning, respecting, doing, participating. Thinking deeply about this may require fine-tuning your approach to witnessing or participating. Or it may require fine-tuning your question to account for social boundaries, power dynamics, your personal concerns, and other ethical considerations tied to the processes of being ethically present and involved in people's lives. You may also find that, while in the field, these questions may radically change to meet the fluctuating circumstances of field research.

For this exercise, you don't have to import your RD, MO, and scoping question again, but if you want to, please do. Example 9.2 about Rahman's project illustrates how he planned to strategize the direct observation of real estate transactions in the private and public spheres. His project suggests that there could be two interacting questions for his observation activities. Also, notice that the interactive questions collectively contribute to answering the data-gathering question.

Typically we find that having two open but focused interacting questions per activity, which should be clearly oriented toward witnessing and/or participating, helps you focus in ways that make data gathering effective.

EXAMPLE 9.2 Rahman's Interacting Questions: Observations

Data-Gathering Question: How do plots become liquid, or financialized, forms of land?		

Process Cluster Name: Financialization		

Data Sets	Interactivities and Methods	Interactive Questions: Observations
Get data about real estate transaction processes among real estate agents, purchasers, sellers, and bureaucrats	Participant observation in real estate transaction processes in private offices in Lahore and in villages where plots are sold	Who is capitalizing and financializing plots as real estate and how are they learning it? How does the process work? Whose land is being sold and what impacts does that have on families and villages?
Understand how people learn about buying plots and how and why they strategize investments	Participant observation among Pakistani diaspora WhatsApp chat groups exchanging information about real estate plots in Lahore	How and why do investors engage in digital technologies for real estate transactions? How does the exchange of information impact buying and selling patterns?
Get information about how paper documents are changing to digitized transactions that are tied to processes	Observation in real estate offices and public government documents offices in Lahore	How do bureaucrats process, use, and disseminate digitized real estate transaction data and what effects does it have on real estate activities writ large?

1. Consider these prompts before drafting your questions.

   a. One way to think about social experiences and processes is to imagine them as the 4Ps in action and in context. What do you want to know about these actions to answer your data gathering questions?

   b. Ethnographers want to have the opportunity to respectfully, and with permission, engage with social, cultural, and environmental processes. What kinds of questions will help you engage the "hows" of the processes you are interested in?

Fill out Table 9.3 for the questions and methods you have planned for each observational interactivities.

TABLE 9.3 Interacting Questions: Observations

Data-Gathering Question:		
Process Cluster Name:		
Data Sets	**Interactivities and Methods**	**Interactive Questions: Observations**
Add rows as needed...		

EXERCISE 9C **Compose Questions for Your Conversational Interactivities**

You're ready to assess your interlocutor group selections and draft a small set of questions to ask all of them and specific groups. In this exercise, you will: (1) describe subject groups, their characteristics, and the logistics and ethics of communicating with them, and (2) develop some common and tailored interacting questions for these subject groups. A small group of questions is a way to focus on what a congruent approach to question format development looks like within a multidimensional design process.

Additionally, the exercise offers an optional prompt that generates a more extensive "interview instrument," for one or more subject groups. The interview instrument contains the entire list of questions that you will ask your interviewees during an interview. This list can be as few as ten questions or as many as thirty-five or forty. You will need an interview instrument for formal approvals of various kinds, such as working within institutions or within self-governed communities, so you may consider this iteration your first draft.

The reflection on interlocutor groups will help you articulate the broad and specific demographics and characteristics of groups associated with all the interactivities that you listed in Table 9.2. These are people you will interview or converse with. In some cases they are associated with what you wish to observe or participate in and sometimes not. But taking time to note their unique positionalities and contextual conditions will help orient to the process of asking questions. Here are some prompts for your reflection:

1. *Make notes about each interlocutor group you wish to speak with.*

    a. Write about how you expect to connect with and meet them.

    b. Write about where and when (time of year, time of day, etc.) you will do the interviews and have conversations (during/after work). Will this be in real life and/or online? In homes or workplaces? How will you ensure people's comfort and confidentiality? If you don't know, speculate on possibilities.

    c. Address the relationship between participant observation and interviewing. Will you observe and interview? What's the process for doing this and why?

d.   Address any problems of access and conversational issues you might encounter.

e.   Take time to reflect on language and other communication issues as well.

Now comes the time to draft four to five guiding interview questions (see Table 9.4). Remember that these need to reflect the data sets and contribute to answering the data-gathering questions; they also need to be responsive to the specifics of each group you want to encounter.

Example 9.3 lists the questions that can be asked of interlocutors in Ellen Kladky's project, in order to answer the data-gathering question about her process cluster, *class*. For now, just appreciate the correspondences between the interview questions and the data-gathering questions—how will the interview questions address the data-gathering question regarding the way people enact the concept of class? Notice that there is a wondering about which method might be needed to gather data about "personal responsibility."

Here are some prompts to help you draft some questions for three kinds of interlocutor groupings. You may modify these questions later or add to the number you have in the last prompt. For now this activity will break the ice on the process of composing interview questions.

TABLE 9.4 Kladky's Interacting Questions: Interviews

Process Cluster Name:	
Data Sets Connected to the Process Cluster:	
Data-Gathering Question:	
Interviewee Groups	**Interview Questions**
Add rows as needed . . .	

EXAMPLE 9.3 Kladky's Interacting Questions: Interviews

Process Cluster Name:
Class

Data Sets:
- Learn about family generational histories—from secure industrial jobs to insecure work over time; and how families understand class and their class position
- Understand the who, what, aims, and logics of Christian financial literacy programs
- Understand how personal responsibility emerges and in what financial contexts
- Understand affective, racial, and cultural identifications with money and finance

Data-Gathering Question:
How do white families reimagine and enact the concept of class, as households over time have transitioned from relatively secure livelihoods to living primarily with debt?

Interviewee Groups	Interview Questions
Interviews and life history interviews with individual family members across generations	Please tell me about your experience working in the mining industry. What has life been like for you and your family since you got laid off? How do you strategize financially supporting yourself and your family? What is the difference between poverty, working class, and middle class for you? How do you identify with these terms?
Interviews with workers and organizers of Christian financial literacy programs	How did Christian financial planning get started and why is it so prominent here in Appalachia? Who are the people that typically come to these workshops and why? How do people describe to you their relationships to money and finance? Can you give me some examples of how the workshops are successful or not? *Need a method to understand how personal responsibility is taught in workshops and embodied (or not) in participants*

2. *Draft four to five questions for each interlocutor group.* Make sure that your questions pertain directly to each of your data sets and data-gathering question themes. Most likely your questions to will differ across interlocutor subject groups. Place your responses in Table 9.4.

3. *Draft four to five in-common questions for multiple or all subject groups.* Among the interactive questions that you've constructed for each interlocutor group, are there any questions that could be posed to multiple or all groups? If you're not sure, refer back to your process clusters and data sets in order to consider themes in common to some or all of the groups you will be interacting with. You can insert those questions and place an asterisk next to them indicating questions in common.

    If you wish, you can now create a separate interview instrument drawing on the questions you've already constructed. Keeping it to twenty to thirty questions is a good start. You can follow the last prompts. But we also suggest consulting methods handbooks for more details.

4. *Make sure your questions are respectful and logical for the people you are engaging.* It's important to consider that you may be meeting the interviewee for the first time. If so, you need to start with biography or oral history questions (professional experience, educational background, etc.), which are highly important for narrating data in project write-ups. Moreover, you might want to start with "What's your activist/educational background?" or "How did you get interested/involved in Black Lives Matter?" before you ask, "How do you feel about the police department?"

5. *Allow questions to logically build on each other.* Creating a good sequence helps construct more and more knowledge data with each successive question. It also builds rapport with interlocutors. Be sure not to run several questions together—interviewees will usually choose and answer only one. Consider this a first draft of questions you'll need for Ethics Review Board approval.

You now have a strong baseline for interviews across groups for each of the data sets you need. Next, we'll turn to archives, which are best approached with questions in mind rather than without.

EXERCISE 9D **Account for the Archive**

By *archive*, we are referring to a range of formal and informal resources. These can be research archives and special collections located in libraries; these could also be gray literature, literature produced by interlocutors, industry or organization literature, family archives, diaries and paraphernalia, online literature, archived social media, histories found online, film and other archival media, and embodied practices of documentation and memory, and so on. Use these prompts to write notes about what kinds of archives you'll be accessing.

1.   *Write about what counts as "the archive" in your research project.* Be specific about the spaces, access issues, languages, and kinds (research libraries and other collections, cached internet material, etc.). Find resources to understand the politics of archiving as an act of power, as an act denied certain groups, and as a form that social groups may or may not have a sovereign relationship with.

2.   *Write about the structure of the archive and how it fits into a larger archive or group of archives.* Think about who runs it, how it runs, who uses it, how they use it. If not now, then in the future consider spatial organization, location of the archive, access, and so on.

3.   *Write about the kinds of storytelling or other engagements used to represent your data.* Examples include family sagas, life histories, images, gesture, orality/literacy, literature/fiction, origin tales, still or moving images, sound, senses, performance, memory, art, expression, multiple medias, the enduring, the ephemeral, or multinodality. Pick one (or more) that requires both ethnographic and archival data and discuss.

Now that you've brainstormed these prompts, you will construct a table where you can see the congruence between archival interacting questions, data sets, and data-gathering questions (see Table 9.5). All data sets,

Process Cluster Name:	Data-Gathering Questions	Archives	Interacting Questions
Cultural Memory Learn about changing relationships of racialized Black bodies to metal as well as disciplinary technologies and tactics over time Get data on how slavery constructed and delineated racialized Jamaican bodies and identities within plantation geographies Understand cultural and sensory memories and practices about security	How do certain contemporary social practices and materials speak to the connection between the conception of security in the colonial past and in the postcolonial present?	The National Library of Jamaica, the Jamaica Archives and Records Department The Main Library at the University of the West Indies (UWI) Library of the Jamaica Constabulary Force (JCF) Former plantations and museums in the Jamaican parishes of Kingston, St. James, St. Elizabeth, and St. Catherine that contain metal disciplinary and torture instruments	How is Kingston as a historically urban and Black city, spatially and ontologically constituted through historical uses of metal material artifacts? What are the colonial histories of crime and violence in Jamaica that helps to genealogically situate contemporary social politics of the secure home and communities?

TABLE 9.5 Data Sets–Data-Gathering Questions–Archives Relationship

Process Cluster Name and Sets:	Data-Gathering Questions	Archives	Interacting Questions
Add rows as needed . . .			

data-gathering questions, and interacting questions that you will pose to the archive should all be linked. In Example 9.4, notice that when each interacting question is answered, it collectively contributes to answering the data-gathering question. This is made possible via the direct connection and congruence between the data sets and the interacting questions.

4. Fill out Table 9.5 with the archives you plan to engage and draft three to four interacting questions for each kind of archive.

EXERCISE 9E **Conceptualize Nonstandard and Other Methods**

Remember that in Example 9.3 there was a placeholder: *Need a method to understand how personal responsibility is taught in workshops and embodied (or not) in participants.* It's good to indicate when disciplinarily standard methods don't seem right for the kinds of information you need to collect.

Several kinds of methods are considered standard to the institutionally formal discipline of ethnography, including linguistic analysis, quantitative data collection, survey research, focus groups, life history, semistructured and other kinds of interviews, and direct/indirect/participant observation. Moreover, emergent methods are being used in many domains, such as digital and virtual worlds research, visual ethnography, and community-based research. There are also less frequently used, or at least less emphasized and nonstandard, ethnographic methods, including autoethnography, sensory anthropology, and drawing, among others.

In multidimensional research design processes, you can also create customized methods to help you and those you work with add dimension to the project effectively. One ongoing aspect of research life is to stay abreast of innovative methods, which are often being produced in response to new ideas about how to conduct collaborative, ethical, and socially responsive investigations.

For example, a researcher we worked with, Colin McLaughlin-Alcock, conducted an ethnographic project in Amman, Jordan, about public art. After getting to know the artist community quite well via participant observation and interviews, he decided to create an online descriptive catalog of artists in the city. After he launched the website, he often got stopped in the street or in cafés and was told what he got right and wrong about the entire endeavor. McLaughlin-Alcock told us that he would never have obtained the kind of information he did in those conversations if he had stuck only to participant observation and interviews.

In another example, an undergraduate student studied the culture of underground drag-car racing in the Los Angeles Latinx community. He found that his interlocutors had surprising things to say about the sensorium of drag racing. And so he added an additional method. He placed his smartphone on the dashboard of a car and video-recorded the races; he played them back to the drivers and passengers and evoked some incredible conversations that would not have happened if he had only done the participant observation or interviews. He created a film from that raw video footage as part of his final honors thesis project paper, which added a sensorial dimension to the written description of the research project.

Other methods can add conceptual dimension to a project, such as developing three to five highly pertinent yet narrowly constructed questions for a focus group, such that a dialogue among interlocutors can bring surprising results. The same is true for quantitative surveys that attempt to configure a research context by talking to many people about broad-stroke, rather than narrow, contexts.

1. *Consider nonstandard or other methods for your project design.* What sorts of other or nonstandard methods might be important for your project: surveys, oral histories, drawing, focus groups, and so forth? How and why might these methods be ideal for data-gathering? Record your thoughts in Table 9.6. (For these and more traditional methods, you can consult an array of very fine books on anthropological methods.).

TABLE 9.6 Other Methods

Specific Data to Be Gathered	Other Methods to be Used	Why Might These Methods Work Better Than Others?
Add rows as needed . . .		

Now that you have created preliminary questions for participant observation, interviews and conversations, the archive, and other methods, it's time to assess them.

1. *Read and compare tables.* Gather together and look at all the interactivities you've constructed.

2. *Create a single table linking connecting and interactivity zone elements.* Use your previous tables to fill out Table 9.7.

3. *Do a congruence reflection.* This reflection develops your research design muscle and sensibility—the capacity to consider what you have, both the correspondence and the gaps. Eventually you won't have to do these reflections because you'll know, sense, and intuit the need for correspondence and overall project congruence across these zones. Doing so will support your capacity to multidimension; this can be achieved by showing you where

TABLE 9.7 Checking for Congruence Gaps between Connecting and Interacting Zones

RD:				
MO:				
Scoping Question:				
Process Cluster Name + Data Sets	**Data-Gathering Question**	**Interacting Questions**	**Which 4Ps or archive?**	**Interactivities and Methods**
Add rows as needed . . .				

opportune gaps and connections need to be made in the various zones that will be innovative and yield significance.

a.   *Compare all the interacting questions that you've linked to the data sets and their methods strategies.* Check for *congruence.* This means that all the questions you're asking in the field (interacting) will contribute to answering one or some of your data-gathering questions. Consider whether questions, like biographical questions ("Where were you born?") or questions about when and how an archive was created ("Who collected these papers?"), correspond to other interacting questions that will answer the data-gathering questions. It also means that the data-gathering questions you developed in Module 8 have corresponding interacting questions so that they can and will be answered.

b.   *Do you see any gaps or find any problems with correspondence?* If you see any gaps, that may mean you need to add or adjust the interacting questions or methods strategies so that they are congruent with the data sets and data-gathering questions.

c.   *Do any gaps point to a need to rethink your project elements or your multidimensional work to include a literature, a 4P, or something else that was not clear to you until now?* If it's a big gap, then it may direct you just to the scoping zone or scoping question, therefore even creating an opportunity to rethink the data-gathering questions, the RD and the MO.

d.   *If you are having trouble recognizing congruence or gaps, answer the following question:* "How do your interacting questions and methods strategies reflect all the data you need to collect as specified by your data sets and your data-gathering questions?"

Know that there is no expectation for everything to match up exactly. However, free-writing a response to this question may help reveal the tensegrity between the connecting and interacting zones, and it may help you know what's clear, what's fuzzy, and what needs time to bake.

Now you can fill out additional grid elements. For each process cluster row in the bottom part of the grid of Table 9.8, list the specific 4Ps that correspond with the cluster. Add the (usually multiple) methods that will be used to answer each data-gathering question. You can draw on your Table 9.7 work, which shows how different methods are used to answer one data-gathering question.

If you prefer, you may remove the 4Ps box in the upper half of the grid in your final iteration. This is because the 4Ps are now linked to corresponding process clusters, data sets, data-gathering questions, and methods in the lower half of the grid. Moreover, broad contexts can be located in the RD, scoping questions, and significance statement.

TABLE 9.8 Final Project Grid

Research Description
Multidimensional Object
Scoping Question
Significance Statement (reduce to 50 words)
Key Concepts **4Ps:** **Broad Contexts:** **Other Concepts:**

Process Clusters	Data Sets	Data-Gathering Questions	Specific 4Ps *Who/what/where will you engage to answer this question?*	Methods *How will you get the data?*
Add rows as needed . . .				

The simple but sophisticated Final Project Grid table that you have just completed will guide you well into the future. Unlike a proposal narrative, within which project elements become lost within a dense multipage composition, the grid is a lucid meshwork of elements that show up distinctly and clearly as they play in concert. The grid keeps you closely attuned to your project, allowing you to fine-tune it as necessary. Going forward, it helps you harmonize your aims with your research experience as the project goes through its next phases and inevitable changes. For inspiration, see appendix ii, which showcases Annie Wilkinson's partially filled-out research project grid.

Collective Concept Workspace 9

Congratulations! You completed all the major concept work in this handbook! You've done the heavy lifting of developing strong relationships between the three zones of inquiry. This is not easy work, so no matter where you are in the process, the accomplishment here is completion. The work of learning more and adjusting this foundational ground will come later.

The group work has two aims: review this module's exercises for congruence between the different project zones, and participate in an intentional closing (or continuation) of group work. Consider revisiting your agreements before you begin.

Concept Work

1. As a group, first check in and see how it feels for each member to arrive at this point. When you look at the different categories of the final project grid, how does each box feel in the body? Mark what feels "right" and what feels "not quite there" or "off." Then discuss.

2. Now that you have established a broad and thorough conceptual rendering of the project, do the methods map to your broad conceptual inquiry? Talk about how you see the methods-concept relationship working in your project design thus far.

3. Review the interacting questions and data for interviews, the archive, participant observation, and nonstandard/other methods. Does everything match up with data sets and data-gathering

questions? Is there more emphasis in one area and less emphasis in another? Is anything missing? Discuss any holes or challenges with your group.

4. Discuss the interacting questions for interview instruments if you completed them. Supportively evaluate how the questions flow and if their intentions for gathering specific data are clear. You can also practice doing role plays with your interview instrument.

5. Developing consonance across all three zones is the most important and often the most difficult part of this process. The objective is to make sure that the interacting zone directly informs the connecting zone and the connecting zone directly informs the scoping zone. Discuss in your groups how the congruence across zones is working.

6. Use the different project grid categories to assess your work. Do the literature, 4Ps, methods, and different zones of questions all connect and provide an interrelational feel? Where is there consonance, and where is there confusion? What's missing? What's needed now?

Closing or Continuing the Collective Concept
Workspace

Your final project grid represents a rite of passage from intensive concept work into an actual living and breathing project; it is essentially the beginning commitment to a long-term relationship. Your group has constituted at least one support system that has helped you arrive at this point. Now the question is, what role does your group play in helping you mobilize the grid via proposal construction, field research, data analysis, and ethnographic writing or other modes of research representation and engagement? Whether your group decides that the project grid marks the end of group work or has plans to continue to work together over the long term, then consider marking this important rite of passage:

1. Share how far you've come from the first construction of a research imaginary to the final project grid.

2. Share appreciations for working together through all the modules.

3.   State how you will commit to rest and pause before moving on to the next stages of project design.

4.   If continuing to work together, begin to share ideas on future concept workspace configurations.

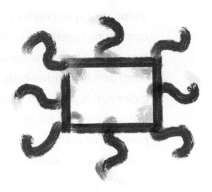

Mobilize Your Research Project Grid

Congratulations! You have designed the framework for an ethnographic project. Allow time to reflect on how far you have come: from sketching a research imaginary to completing your final research project grid. The grid clearly provides, at a glance, the project's main purpose and central questions, conceptual and empirical links, literature engagements, potential significance, and basic methods plan. Think of the research grid as a project representation and communication device with many additional applications.

Previous modules contain exercises to develop a particular goal, such as crafting an RD or a scoping question. In contrast, this module helps you put your research project grid into action.

MODULE PURPOSE
Use your project grid to prepare ethics board applications, write grant proposals, or do multidimensioning during field research.

All the exercises in this module can be treated as stand-alone concept work for very particular project goals. After reading them, you can make two workflow decisions. First, you can decide how to schedule your next steps, including any necessary pauses for yourself and with your group if you are continuing together. Second, you can decide which exercises suit your current project implementation needs. If you need to check for overall project congruence (in either early or later project design stages), then start with Exercise 10a. If you need to apply for permission from an Ethics Review Board to conduct your project, then go to Exercise 10b. If you are ready to begin proposal or grant writing, then we encourage you to move to Exercises 10c through 10e. Those exercises address three ubiquitous proposal sections: project aims/problem/description, the literature review, and your methodology. Lastly, we provide guidance on how to carry multidimensional concept work into the field research stage (Exercise 10f).

Before going further, it's important to cultivate a rested body and open mind so that you can assess, finalize, and engage your project with fresh clarity. In a friendly way, pay attention to your immediate needs to pause or process. For example, ascertain any necessary (and possible) breaks that you could take—either days or weeks—before beginning what you need to do next. In addition, take time to contemplate your grid and each of its elements. What does it need, if anything, before you move ahead? If you need to return to earlier modules to do a grid adjustment and multidimensional congruence check, we provide some guidance to help. With these invitations in mind, we suggest that you begin by just reading through the module.

Exercises

EXERCISE 10A Make a Whole-Project Congruence Check

In Module 9, you did a congruence check between the connecting-zone and interacting-zone elements. It's now time to do a congruence check of your whole project, which amounts to a grid composition check as well. In the following prompts, we invite you to pair up grid elements to make sure there is congruence between them as well as within and across the inquiry zones; you may recognize some of these reflection prompts from earlier modules. If you find congruence in all these element-pairing checks, you can rest assured that your grid is in good integral shape.

Respond to the following:

1. *Pair your RD and MO with fresh eyes.* Does your MO still connect and contain the RD elements? If not, make adjustments.

2. *Pair your RD with the scoping question.* Does the scoping question look as if it could be the last line of an expanded RD, such that it could fit into this sentence: "Therefore, this project asks [fill in scoping question here]." Consider adjusting the scoping question accordingly. After you do this, ask whether your MO still overarches the elements and intentions of the scoping question. If not, you may have to go back to your RD and tweak it a bit.

3. *Pair your MO with the process cluster concepts.* Does your MO connect with and feel like it meaningfully conveys all the terms in the clusters? If you feel a lack of congruence, ask yourself if it is consequential or not. If it feels consequential, then adjust for congruence across elements. When you complete this check, adjust the data-set names as needed to make sure the MO connects and contains them.

4. *Pair your MO with the literatures.* Does the MO feel as if it creates a connection among the literature categories you have listed? This connection doesn't have to be overtly direct, but you shouldn't feel as if one of the literatures is completely unrelated to the scoping spirit of the MO. If not, it is time to consider whether the literature categories need to be refined.

5. *Assess the grid's overall congruence.* Come back to the interacting-zone and connecting-zone work that you completed—the 4Ps, data sets, data-gathering questions, process clusters, and methods. How congruent are these with elements of the scoping zone—the scoping question and significance statement—as well as your literatures? Does the MO feel like it holds it all together?

This congruence check can be done any time you make a new version of your project grid.

Develop an Ethics Review Board Application Plan

Institutions that house researchers, or that allow researchers to conduct research in them, usually require every project to undergo an ethics review. In many countries and sovereign spaces the body that oversees this process is called an Ethics or Institutional Review Board, which regulates human subjects research. Each institution or community governing body has its own guidelines for who needs ethics approval and for how extensive that approval process will be (throughout the United States, institutional review includes levels of exemption, expedited partial review, or full review). Many ethnographers must apply for multiple ethics approvals, from their own and other governing bodies. Familiarize yourself with such community and institutional requirements before you begin.

With your completed grid in hand, you can focus on how to address the broad and specific ethical, moral, and justice concerns that will impact you, your research, and the lives of those you plan to work with. It is likely that you have been engaging in ethical considerations all along, as guided by your work group and advisers. If you haven't, or need to do more work, the grid can be a useful communication aid. Consider using it to work with communities you are planning to visit in order to discuss and respond to their ethical concerns and requirements. These can include securing individual and community permissions to do work within social spaces, agreeing to confidentiality and anonymity expectations, and working through any personal and interlocutor safety considerations. There are many texts that address these issues and support your aims to implement necessary processes to ensure just fieldwork practices.

In this exercise, we show how the grid can help you to address vital ethical questions about the relationship between the project's described purpose, its enumeration of participant groups, and its selected methods.

1. *Justify the project's intellectual and social purpose.* When an ethics discussion or protocol form requires you to state and justify the study's "purpose," you can combine your RD and significance statement to answer what you will do and why it is intellectually and socially significant. You can supplement this combined statement by detailing the processes found in your 4Ps as well

as the data-gathering questions, which will help you to bridge the purpose and conceptual framework with what you are going to do as you interact with people and beings. If you are asked for a hypothesis, you can draw on your scoping question and data-gathering questions to develop one or more hypotheses.

2. *Link the purpose to methods.* If you are asked how your data-collection plan is correlated with the project's aims, then draw on the congruence you have established between data sets, data-gathering questions, and their specifically linked methods. That is, use the direct relationship between these to explain how your plans to interact with people and beings will directly address the purpose of the study. You will also have to demonstrate that your participant recruitment plan is being conducted in accordance with the sovereign needs and requirements of the communities you will engage; and as we indicated in Module 9, that's a process you must engage in parallel with this project design process.

3. *Compose a short nontechnical description for the participation consent process.* More than likely, you will need to ask your interlocutors for informed consent when doing interviews or other ethnographic work. The required content of a consent form varies by institution: in fact, many institutions provide document templates for informed consent. These forms often begin with a project description before they enumerate participant rights. For this, you can revise your RD and significance statement in a language and form that specifically address your interlocutors. You can return to the twenty-five-word project description for nonacademic audiences that you wrote in Module 5 to inspire your composition process. Make sure that your study's purpose is clearly explained, including why you wish to recruit particular persons (who have experience, expertise, situated knowledge, etc.).

Writing Grant and Thesis Proposals

Project proposals take different forms, but they all require researchers to detail the project elements in writing. You might have to write complex narratives with sequentially ordered paragraphs. Or you might need to compose one or two paragraph responses to discrete questions such as "What are your methods?," "How will your project contribute to the discipline?," or "What is the social significance of your project?" Either way, grants and thesis proposals have common focal subsections, three of which are nearly ubiquitous: the project description, literature review, and methods sections.

The three common proposal sections have particular objectives. The project description summarizes the purpose, aims, scope, and significance of the project and poses its scoping and data-collection questions. The literature review demonstrates how your project engages, extends, and amends relevant bodies of knowledge. The methods section is an overview of where, when, and how you are going to interact with people and gather data.

In all successful proposals, each of these sections connect, conceptually; that is, they demonstrate congruence. For example, the key concepts and processes you describe in your project description section must connect to the studies you cite in the literature reviews. All the activities and processes that you claim that you will investigate must be accounted for in the methods section. In addition, you must demonstrate that the data you collect, using the methods you have chosen, will significantly contribute to answering the project description section's overall (scoping) question.

The good news is that your grid has all you need to write these subsections and maintain their congruence. We will demonstrate how to create a congruent proposal by reverse-engineering grant-winning examples throughout the exercises.

EXERCISE 10C The Opening Paragraph of the Project Description

The opening paragraph of most project descriptions is crucial: like a good novel or news article, it provides a "hook" that draws the reader into your project. To be effective, the description must highlight the project's compelling focus, breadth, and significance. The remaining paragraphs of the description section—which we do not cover here—typically contextualize and situate the project, elaborate its sites and methodological processes, and declare its scoping question, data-gathering questions, and significance.

The opening paragraph, then, is engaging, informative, convincing, and—importantly—short. A short, powerful, pithy statement of the overarching focus and aims of the project (rather than one that belabors points) combined with a concise statement that connects the ethnographic terrain with the scope of the project will hold rather than overwhelm your reviewer's attention. All in all, your opening paragraph should have roughly six to ten powerful and well-organized sentences. To compose this paragraph, you can expand your RD in a way that specifies details about the 4Ps, foregrounds the multidimensional object, states or hints at significance, and sets the stage for, or specifically poses, a scoping question.

Examples 10.1 and 10.2, from Sean Mallin's and Tariq Rahman's projects, break down versions of opening proposal paragraphs sentence by sentence, explain their communicative function, and show how those sentences map onto grid elements.

EXAMPLE 10.1 Reverse-Engineering the Opening Paragraph of Mallin's Project Description

Opening Paragraph: Six years after Hurricane Katrina, one in four homes in New Orleans remains vacant. These vacant properties not only represent the former residences of thousands displaced by the disaster, but have increasingly been framed as a social and economic "problem" stalling poststorm recovery. In 2010, the mayor of New Orleans launched an aggressive campaign against "blight" in the city, targeting these vacant properties through the development of a comprehensive code enforcement process. This process, which includes inspections, hearings, and the possible seizure and sale of blighted properties, did not exist in a coherent form prior to the disaster. As the city continues to rebuild, the emergence and use of the term "blight" is part of a broad shift in how vacant properties might be framed in relation to Hurricane Katrina, from a discourse of mourning a loss to a discourse of "responsible" ownership and commitment to the city's recovery. The scale at which code enforcement is being pursued in New Orleans—as the primary redevelopment strategy for more than 45,000 vacant properties—is unprecedented. This makes the study of post-storm code enforcement both novel and important, because it indicates how the city might respond to future climate disasters, urban planning, and racialized economies of dispossession.

Project MO: Coding blight

RD: Six years after Hurricane Katrina, one in four homes in New Orleans remains vacant. In 2010, the mayor of New Orleans launched an aggressive campaign against "blight" in the city, targeting these vacated properties through the development of a comprehensive code enforcement process—a process that, surprisingly, did not exist before the storm. As the city continues to rebuild, the emergence and use of the term "blight" is part of a larger shift in the continued racialized processes of dispossession with conjoined health, legal, economic, and weather-related dimensions.

Scoping Question: How is the post-Katrina enforcement of "blight" codes in New Orleans conjoining processes of racialized spatial, legal, climate-related, and economic dispossession?

Opening Paragraph Sentences	Sentence Function and Use of Grid Elements
Sentence 1: Six years after Hurricane Katrina, one in four homes in New Orleans remains vacant. *Sentence 2:* These vacant properties not only represent the former, mostly Black-owned and occupied, residences of thousands displaced by the disaster, but have increasingly been framed as a social and economic "problem" stalling poststorm recovery.	*Sentence 1* distills RD keywords into an attention-focusing statement that conveys an urgent problem. In this case, the author invokes a compelling and shocking image: a Southern US city that is now 25% vacated. This makes the reader immediately want to ask: Whose homes are these? Why are some people unable to return but others are able to? What is happening to New Orleans' people and places in the hurricane aftermath? These questions set the stage for a scoping question—either here in this opening statement or in the next paragraph. *Sentence 2* immediately contextualizes the opening statement regarding who has been displaced but also conceptually connects the who of property vacancy to the main discursive and actionable city planning shifts that underscore urban recovery.

Sentence 3: In 2010, the mayor of New Orleans launched an aggressive campaign against "blight" in the city, targeting these vacant properties through the development of a comprehensive code enforcement process. *Sentence 4*: This process, which includes inspections, hearings, and the possible seizure and sale of blighted properties, did not exist in a coherent form prior to the disaster.	*Sentence 3* shifts out of compelling opening statements and introduces the main ethnographic focus, which also comprises a term from the MO: blight. The second clause of the sentence emphasizes a key focus of the MO: the city justifies targeting properties by drawing on the discourse of blight. *Sentence 4* focuses on the processes pertaining to code enforcement, revealing that such aggressive targeting did not exist prior to Hurricane Katrina, making one wonder, why now?
Sentence 5: As the city continues to rebuild, the emergence and use of the term "blight" is part of a broad shift in how vacant properties might be framed in relation to Hurricane Katrina, from a discourse of mourning a loss to a discourse of "responsible" ownership and commitment to the city's recovery.	*Sentence 5* explains the social significance pertaining to a shifting discourse that the city needs so that blight and code enforcement can be facilitated as a cornerstone for urban recovery and development.
Sentence 6: The scale at which code enforcement is being pursued in New Orleans—as the primary redevelopment strategy for more than 45,000 vacant properties—is unprecedented. *Sentence 7*: This makes the study of post-storm code enforcement both novel and important, because it indicates how the city might respond to future climate disasters, urban planning, and racialized economies of dispossession.	*Sentence 6* highlights the scale of massive dispossession that many face and underscores its novelty in the city's urban development. *Sentence 7* not only connects the MO to broader urgent and changing urban patterns (the scoping terms), but makes them relevant to cities facing similar compounding problems. These connections show that the project has broader significance: it will focus and amplify theorizations of how legal and economic power interacts with disaster forces to remake forms of property, spatial inequalities, and experiences of place and belonging.

EXAMPLE 10.2 Reverse-Engineering the Opening Paragraph of
Rahman's Project Description

Opening Paragraph: In the city of Lahore, Pakistan, plots of land are increasingly traded like stocks. Land has traditionally functioned as inherited property or a long-term investment. But across a diasporic market, plots of land circulate between local residents and overseas Pakistanis located in Dubai, Malaysia, England, the United States, and elsewhere; and in Lahore's inner-city the state is digitizing centuries-old land records in order to increase the liquidity of local assets. These changes reflect a long-term outcome of the war on terror such that diasporic Pakistanis, feeling uncertain about their future in the West, transferred a mass of wealth to buying land. Under these conditions, land was a stable but largely obstructed asset, and therefore the financialization of land relied on dense webs of social relations alongside the transnational proliferation of financial rationalities and communication technologies. The outcome is that value has gotten attached to speculative exchange rather than land itself. Because financial markets favor abstract forms of value, land in Lahore is not only physically embedded but also made liquid within multiple value regimes. When complete, this project will demonstrate how local histories and practices entangle with the emergence of modern markets in the Global South and the reshaping of postcolonial cities under financialization, military imperialism, and migration.

Project MO: Liquid land

Scoping Question: How do emerging techniques to buy and sell land in Lahore shed light on the ways that Pakistani social life is being transformed by urban responses to the dynamics of migration, global finance, and imperialism?

RD: In the city of Lahore, Pakistan, plots of land are increasingly traded like stocks. Though land has traditionally functioned as inherited property or a long-term investment, since the early 2000s it has become liquid, or a good that can be quickly bought and sold as prices change. Across a diasporic market, plots of land circulate between local residents and overseas. While financial markets favor abstraction and rational economic decision-making, land in Lahore is not only physically embedded but also enmeshed in multiple and overlapping regimes of value and legality. Pakistanis located in Dubai, Malaysia, England, the United States, and elsewhere; and in Lahore's inner-city the state is digitizing centuries-old land records in order to increase the liquidity of local assets and respond to the incursion of capitalist and imperialist forces. This project investigates the relationship between the changing qualities of Pakistani land and the transformation of Pakistani social life by migration, global financialization, and military imperialism.

Sentences	Sentence Function and Use of Grid Elements
Sentence 1: In the city of Lahore, Pakistan, plots of land are increasingly traded like stocks.	*Sentence 1* is an attention-focusing statement that immediately highlights the power of the MO. It's notable because the anthropological literature on finance does not focus so much on land, much less on financialization processes in South Asia.
Sentence 2: Land has traditionally functioned as inherited property or a long-term investment. *Sentence 3*: But across a diasporic market, plots of land circulate between local residents and overseas Pakistanis located in Dubai, Malaysia, England, the United States, and elsewhere; and in Lahore's inner-city the state is digitizing centuries-old land records in order to increase the liquidity of local assets.	*Sentence 2* shows that previous forms of land value dominated real estate, indicating that the process of financialization is relatively new. *Sentence 3* shows how the 4Ps of the project are driving changes in land's valuation.
Sentence 4: These changes reflect a long-term outcome of the war on terror such that diasporic Pakistanis, feeling uncertain about their future in the West, transferred a mass of wealth to buying land. *Sentence 5*: Under these conditions, land was a stable but a largely obstructed asset, and therefore the financialization of land relied on dense webs of social relations alongside the transnational proliferation of financial rationalities and communication technologies. *Sentence 6*: The outcome is that value has gotten attached to speculative exchange rather than land itself.	*Sentence 4* and *Sentence 5* establish the connection between the ethnographic focus and its broad context, namely how the war on terror put new avenues in place for financialization of land to thrive: new social and political, and technological structures and relationships. *Sentence 6* underscores the current outcome of these events: land speculation as a primary form of valuation.

Sentence 7: Because financial markets favor abstract forms of value, land in Lahore is not only physically embedded but also made liquid within multiple value regimes.	*Sentence 7* mobilizes MO terms to show how the study of land speculation in Lahore will reveal a major departure in how value is understood in and across markets.
Sentence 8: When complete, this project will demonstrate how local histories and practices entangle with the emergence of modern markets in the Global South and the reshaping of postcolonial cities under financialization, military imperialism, and migration.	*Sentence 8* justifies and demonstrates the project's multidimensional significance, showing how the financialization of land in Lahore is not confined to one place or singular actions. But rather that they reflect emergent forms of financialization that have important impacts on cities and urbanization.

After you study these examples, and their different ways of covering vital introductory ground, we provide you with guidelines to craft your own paragraph. If you follow them, you'll address questions and concerns that all reviewers have right away.

1. *Gather your project grid elements.* Table 10.1 shows the project grid elements that you need to begin—use this as your opening paragraph workspace. Paste your grid elements into each

TABLE 10.1 Grid Elements for a Project Description's Opening Paragraph

MO:	
RD:	
4Ps:	
Scoping question:	
Significance statements:	

row. When composing the opening paragraph, start with your RD and use the rest of the elements to expand it.

2. *Guidelines for Writing a Compelling First Paragraph*

   a. *Encapsulate the project's core focus and purpose* in a specific, evocative, and clear way. You must make sure that your project's description is recognizable as timely, compelling, and "scopable" to other sites and broad contextual social processes.

   b. *Set the ethnographic scene* in a cogent way that supports the project's specific aims and generalizability. This requires specifying your key 4Ps, main social contexts, and the core compelling social problem or issue.

   c. *Introduce at least one MO word to create a congruence term or phrase* that you can deploy throughout the proposal to help it hang together.

   d. *Include key significance phrases and terms* to directly address or gesture to the "So what?" question. Why does this matter—to whom, and where? Here you can include your *scoping question*, which gestures to the significance, stakes, and potential theory-making of the project.

The purpose of each sentence is summarized in Table 10.2.

TABLE 10.2 Possible Structure of a Project Description's Opening Paragraph

First Sentence	Uses RD and/or MO elements to create a pithy statement that conveys an attention-focusing problem.
Next Sentences	Discuss the project's theoretical and empirical focus that includes the 4Ps. They contextualize the opening sentence, introduce main ethnographic focus, show what's new or unaccounted for, and demonstrates social significance.
Final Sentences	Connects the RD and/or MO elements to broader context and scoping terms, signaling disciplinary significance.

An opening paragraph for the project description section of a proposal can be followed by paragraphs providing details on the big-picture contexts, description and justification for the research sites, and data-gathering questions; all of these should reflect back to the significance and stakes represented in the opening paragraph. Nothing starkly new should be addressed in subsequent paragraphs. This is how important the opening paragraph is: it sets the stage for connection of all project elements while, at the same time, demonstrates the multidimensional and theoretical possibilities of the project.

EXERCISE 10D The Literature Review Section

We often find that students think that the purpose of a proposal's literature review section is to report their knowledge of the field. This is not an accurate assumption—in fact, it can jeopardize the success of your proposal. It may even suggest to an adviser or reviewer that you already know the answer to your scoping question (or hypothesis if it is required). Since you are competing for funding or seeking ethical approval for your project, the literature section of a proposal should have one aim: to make a citationally framed argument about why your innovative project needs to be supported.

As was described in Module 2, your effort to intervene in the literatures that have been done on your topic is often referred to as a "gap" in the literature. For grant, thesis, or dissertation proposals, it's not enough to say that you've identified such a gap or that "no one has written about" your interlocutors or the process you're focusing on. You must state *why* this gap is problematic or significant, and you have to state *how* your project's aims and object of study (your MO) will fill or correct this gap, intended to advance scholarly knowledge. In other words, it's an argument that identifies the literature-based reasons for why your project's objectives are necessary. That argument should reveal (and assert to fill) scholarly gaps, convey significance, and enhance disciplinary and social understandings of a process.

This argument can be built by (1) using your key literatures to frame the review, and (2) using your MO to structure your argument about what kind of new work needs to be done in those literature conversations. Remember that both your literatures and your MO have generated project multidimensionality, which helps to open up previously unknown aspects of literature conversations.

In this exercise, we demonstrate how to write the opening paragraph of a proposal's literature section. We also provide a snippet view of how to write about one (of most likely three) literatures that your project will engage. Sometimes writing a literature review, especially one that has a word-length requirement, requires identifying three or four more fine-tuned topical categories than your broad key literatures.

Let's look at an excerpt from Sean Mallin's actual grant (Example 10.3). In the introduction to the literature review, he begins with a paragraph that signals the particular key literature topic he will focus on. These clearly relate to well-understood broader key literatures: law and property (key literature: legal anthropology), urban governance (key literature: urban anthropology), and markets (key literature: economic anthropology). We provide

EXAMPLE 10.3 Mallin's Literature Section Excerpt

Text	Analysis
Opening Paragraph of Literature Section: Through an examination of code enforcement and the meaning of "blight" in post-Katrina New Orleans, this project engages theoretical and empirical debates on law and property, urban governance, and markets, as well as recent area literature covering the aftermath of Hurricane Katrina. In doing so, this project analyzes the constitution of "blighted" properties and "irresponsible" owners; the logic and practice of municipal law and code enforcement; and the qualitative and quantitative processes involved in commodifying vacant properties.	This paragraph introduces the grant's literature section by first highlighting the three key literature topics that the project engages: a) law and property b) urban governance c) markets The paragraph creatively uses a key term from his MO—*blight*—and RD/opening paragraph elements—to frame how the literatures will be engaged. This is a specific multidimensioning practice that connects the literature topics with the project's major 4Ps and conceptual foci. It's generally good to summarize these connections in the beginning of every section of a proposal.

Snippet from One of the Project's Literatures Subsection Heading: "Law and Property"

Sentence 1: Law and property were central concepts in early anthropology (Maine 1861; Malinowski 1926; Gluckman 1965) and have remained important in work across the social sciences (Hann 1999; Ostrom 1990; Radin 1993).

Sentence 2: Recent anthropological studies have focused on changing property relations in colonial and postcolonial contexts (Moore 2005; Shipton and Goheen 1992), in post-Soviet states (Humphrey 2002; Verdery 2003), through new biotechnologies (Strathern 2005), and through the interaction of traditional knowledge and intellectual property (Brown 2003).

Sentence 3: However, few have studied changing property relations in US urban contexts, much less in the aftermath of a large-scale disaster.

Sentences 4, 5, and 6: Of course, property is an ambiguous concept (Verdery and Humphrey 2004). The term carries numerous meanings, values, and associations (Coombe 1998; Feld and Basso 1996; Stark 1996), and is bound up in culturally specific notions of personhood (Pottage and Mundy 2004; Strathern 1999). Moreover, property rights are enforced through a diverse set of practices such as mapping, the construction of fences, and the policing of urban space (Blomley 2004; Mitchell 2002).

Sentence 7: While existing scholarship has examined the politics of public and private property (Low and Smith 2006),

Subsection Heading: "Law and Property" focuses on the broader category of legal anthropology. You may choose to be this broad or narrow in how you represent the category.

Sentence 1 locates "law and property" in both anthropology and social sciences by citing a few sentinel works.

Sentence 2 focuses on more recent anthropological publications that are found in this literature's major concepts, topics, and areas. This demonstrates that the proposal writer sees the project as related to the persistence of colonial property regimes, and sets up the next sentence to indicate how the project differs from what has come before.

Sentence 3 is the literature gap statement, generally stated, which is contextualized by some of the 4Ps.

Sentences 4, 5 and 6 name very specific and key concepts in the "law and property" literature that are near to, or embedded in, the project's RD and its conceptual design.

Sentence 7 summarizes the last two sentences and shows not only a gap in the literature but a potential concern over how understandings of property could be differently conceptualized (out of binary frameworks).

this project shows how a binary understanding of property is insufficient (Blomley 1994; Gibson-Graham 2006).

Sentences 8 and 9: In post-Katrina New Orleans, debates over the relative merits of "public" or "private" property have been overshadowed by debates about the social and moral responsibilities attendant to property ownership. This project will move beyond a binary understanding of property to examine how property articulates with a broad range of concerns in the context of postdisaster New Orleans.

Sentences 8 and 9 provide a justification for the project because it opens up the possibility of understanding property beyond the public/private binary, signaling new multidimensional contexts through which property rendered postdisaster emerges.

his paragraph on law and property, which we analyze to show how a review can mobilize terms from an RD, MO, and significance statements. Note how quickly Mallin identifies the literature gap and then supports its significance.

As you can see, the structure, flow, and tone in this excerpt are highly effective. Mallin asserts project concepts that connect to existing literatures. He then immediately states how the proposed research will depart from or add new questions and dimensions to that literature. The tone of the argument is important. In the example, notice how Mallin consistently establishes novel significance and is clear about how the project will harness and materialize that novelty. Using language like "brand-new," "the question remains...," "this project shows how *x* approach is insufficient," will make new interventions," and "I will extend this work by...," helps reviewers to validate the project's promising contributions.

We recognize that students often face challenges in asserting the importance and multidimensional significance of their work. Try to remember that this process is not about ego showcasing. Being clear about project novelty and significance is actually a helpful and humble approach to proposal writing. That is, it does not assume that your grant reviewers or institutional advisers "should" know your areas of research and the stakes of the project in relation to the literature. Instead, the most convincing and generous style is to demonstrate project significance for a broad community of creatives and scholars. When writing clearly in this demonstrative vein feels challenging, work with someone in your concept-work group to support you through the process. Often our closest colleagues can help us summon the emotional confidence to champion the importance of our projects.

Now you can try it yourself. Table 10.3 contains the grid elements you need to construct your literature-based arguments. After you've filled in the

TABLE 10.3 Grid Elements for the Literature Section

MO:
RD:
4Ps:
Literatures:
Significance statements:

table, draw on the following prompts—they are built on the assumption that you have been engaging a body of readings relevant to each of your key literature categories.

1. *Refer to your key literature to structure the review.* Draw on the literature topics in your grid. Overall, your review must reflect your knowledge of the project's key literature categories, such as medical anthropology or economic anthropology. But reviewers also want you to demonstrate that you understand the particular literature topics your project will contribute to, such as the anthropology of health systems or the anthropology of finance. You will also need to think strategically about how to sequence your literature discussions in a way that emphasizes the multidimensional value of your work.

2. *Create an opening paragraph.* The strongest way to do this is to compose a tantalizingly multidimensional introductory para-

graph for the literature review section. This paragraph must present an argument that the project combines literatures in a unique way. You can use your literatures Venn diagram to visually assist you. To use another example—Forest Haven's project—this paragraph could summarize how the project most obviously contributes to the anthropology of food but also, in a novel and much needed way, to the anthropology of embodiment and the senses and to science and technology studies.

Based on Mallin's example, the structure of an opening paragraph could look like Table 10.4 (but is not limited to this example).

TABLE 10.4 Possible Structure of a Literature Section's Opening Paragraph

Opening Sentences	Draw on the MO to frame and justify the key literatures your project engages.
Detail Sentences	Any further detailed justifications can be provided but they may not be necessary.
Final Sentences	The literatures and MO are linked to the 4Ps, thus justifying engagement with, and expansion of, those literatures.

3. *Consider the literature gap.* Next, strategize how to state, within the body of the literature section, exactly what your project contributes to these particular literature topics. To prepare, refer to your exercise responses in Modules 2 and 4 to engage the scholarly readings you've been doing in order to answer the following questions, which you may do via free-writing or voice-memoing.

a. What old and new research problems, debates, controversies, and theories represent what has defined and is now advancing this literature topic?

b. What has been gained by, and what analyses are insufficient in, current works (from the past five to seven years) that address these old and new problems, debates, controversies, and theories?

c. How does your project address what is being overlooked or what is insufficient?

4. *Develop an argument about how your project will fill the literature gap.*

a. A clear way to organize this argument is to create literature subheadings like "Urban Anthropology" or "Law and Property" that follow the literature review's opening paragraph.

b. For each literature subheading, you can follow Mallin's example of how to build an argument. Specifically, you can use terms from your MO or RD to create congruence in your argument that the project will "enhance knowledge" or "fill a gap."

c. The MO can help you highlight the novel researchable dimensions that other work has not sufficiently attended to. To do this, use through-putting citations that connect to your RD or MO in order to construct sentences that create a flow within each literature topic paragraph. That flow usually works well if it moves from a general overview of what has been done, to a gesture to the specifically vital contributions that your project will make.

d. Practice finding a compelling way to express how your project fills a gap. Following this exercise, provide a concrete example of how you can relate existing work to your project's potential. Here is a possible framework to help you construct your argument:

Literature regarding [your key literature topic] has been concerned with [name an important topic/object x that connects to, but is broader than or tangential to, key terms in your project's specific RD or MO (add in-text citations)]. However, these studies do not address [topic/object *x*'s] relationship to [select terms that relate to key concepts in your RD], leaving open questions about how [topic/object *x*] impacts [name your project's groups, processes, theories, or big-picture contexts]. What is being missed, then, are the important opportunities to understand [refer to elements of your scoping question, data-collection question, and significance]?

You can repeat this formula at the end of each literature subheading.

e. *Concluding a literature section.* Your concluding paragraph presents the opportunity to articulate how the multidimensionality of your literatures opens up new conceptual and theoretical terrain. Drawing on the gaps and the theoretical possibilities of your literature juxtapositions helps to underscore a final argument as to why the research needs to be conducted.

f. *Audit: Proposal literature section dont's.* The practices to avoid below are not written in stone, but they can be considered rules of thumb that help you handle the demands of demonstrating project significance and congruence within a restricted format.

  i. *Don't* write extensively about what is going on at the fieldsite or extensively redescribe what the project is about. This information belongs in the project description. The only time that you might need to include a description is when you're using examples from the field to justify a literature gap. See sentences eight and nine in Example 10.3.

  ii. *Don't* write about how you are going to conduct the project or collect data; this belongs in the methods section.

  iii. *Don't* write an extensive "book report" about how literature "inspires" your project or how others' work is good to "think with." This only justifies your interest in others' research areas and work; it does not demonstrate how your project opens new dimensions in the literature or fills known or unknown gaps.

iv.   *Don't* rely on sentences that simply state that your project "fits into" or "adds to" the literature. Phrases such as "I will engage…" and "I look to…" tend to be vague. They may also indicate that your project is just a new or slightly unique case study that builds on existing scholarly work. They do not convey your project's creative diversion from previous empirical work or its potential for theoretical innovation. Granting agencies may not fund grant proposals that rely on this language because such projects do not arguably demonstrate possible new dimensions in the literature. Yet phrases such as these can be followed by contrasting language, like "however" or "but" to mark a new direction that you're extending in your literature categories.

EXERCISE 10E The Methods Section

A good methods or research plan section ties together the project's conceptual framework and scoping question with explanations about how you are going to gather data. In other words, this section presents an intellectually, socially, and ethically justifiable fieldwork process. A research plan/methods section of a proposal can include the following elements: research sites, data-gathering questions, data-collection methods, data analysis, and research timeline. In this exercise, we provide examples for the opening paragraph of the methods section as well as one data-collection example pertaining to participant observation.

The work you did in Modules 7 through 9 ensured that your fieldwork plans—participant observation, interviewing, surveys, videos, sound, life-history interviews, and so on—are directly tied to your conceptual framework: the key concepts, data-gathering questions, and your needed data sets. This means that a justification for your methodological approach is built into the grid itself. In what follows is just one of several approaches to linking the conceptual aspects of the project directly to the research plan.

1.   *Write an introductory paragraph to the research plan.* The key to
     this paragraph is to interrelate concept work with data gather-
     ing. See Rahman's research plan in Example 10.4.

EXAMPLE 10.4 Opening Paragraph of Rahman's Methods Section

Text	Analysis
Sentence 1: Research for this project will consist of eighteen months of ethnographic fieldwork in the city of Lahore, Pakistan, from August 1, 2019 to January 31, 2021.	*Sentence 1* introduces the research plan with a brief statement about the amount of time that will be spent in the field and the kinds of methods that will be used.
Sentence 2: Through participant observation, semistructured interviews, and archival research, this project will investigate how the social and material qualities of land entangle with processes of financialization in Lahore— becoming essentially liquid.	*Sentence 2* restates the purpose of the project, creating congruence between the methods section and the project description. It does so by linking his MO, liquid land, to proposed methods, which will answer his data-gathering question.
Sentence 3: In doing so, it will be guided by the following data-gathering research questions: (1) What are the geopolitics and histories of land speculation in Lahore? (2) How do plots become liquid, or financialized, forms of land? (3) What are the relationships between land's multiple and overlapping regimes of value in Lahore and emergent forms of global financial speculation? (4) How have new understandings of risk introduced by the war on terror contributed to making land liquid? (5) How are residents of inner-city Lahore responding to, negotiating, and appropriating the remaking of their land as liquid; and how might this transform the topology of the city?	*Sentence 3* connects liquid land to the data-gathering questions that he will bring into participant observation. This part is very important because linking project conceptualization with data-gathering questions shows congruence and provides justification for conducting very particular methods.

Based on Rahman's example, a first paragraph can include the following elements and structure as shown in Table 10.5.

TABLE 10.5 Possible Structure for a Methods Section Opening Paragraph

Opening Sentences	The physical sites, the length of research, and proposed dates
Detail Sentences	Describe the methods to be used and directly connect their justification to an aspect of the MO
Final Sentences	Link how the methods for data collection will answer the project's data-gathering questions

2. *Describe how a particular method will be used to research one of their process cluster types.* As you encountered in Module 9, the data-gathering questions drive and justify the specific methods used to gather data. Example 10.5 shows how a data-gathering

EXAMPLE 10.5 Rahman's Participant-Observation Plan for One Process Cluster

Process Cluster: Financialization	
Data-Gathering Question: How do plots become liquid, or financialized, forms of land?	
Data Sets: Obtain data about real estate transaction processes among real estate agents, purchasers, sellers, and bureaucrats Understand how people learn about buying plots and how and why they strategize investments	*Interactivities and Methods*: Participant observation in real estate transaction processes in private offices in Lahore and in villages where plots are sold Participant observation among Pakistani diaspora WhatsApp chat groups

Get information about how paper documents are changing to digitized transactions that are tied to market finance capital processes	exchanging information about real estate plots in Lahore Observation in real estate offices and public government documents offices in Lahore

Data-Collection Methods (excerpt) *Sentence 1*: Participant Observation: As this project is premised on the idea that financialization emerges between different sites, actors, and technologies in ways that are not always explicit, participant observation will be a vital method. *Sentence 2*: At the developer's office, I will work as an unpaid analyst, a participant in daily discussions about general market trends. *Sentence 3*: Working as an analyst will allow me to observe how market information is discussed, which sources of information are deemed valid, and the ways in which such information is translated into brochures, blogs, spreadsheets, maps, ads, and other formal representations of the market. *Sentence 4*: Working at the office will also provide an opportunity for me to observe the buying and selling of plots. *Sentence 5*: By attending to both verbal and nonverbal interactions between salespeople and investors, I will assess how judgments are made or contested about land and its value. *Sentence 6*: I will also monitor the acquisition of the land from local farmers by the developer. *Sentence 7*: Negotiations between developers and farmers are generally tense, and their breakdown often threatens a project's vitality.	*Sentence 1* restates a key component of the RD: that financialization emerges in many and unexpected ways. This framing is important because it then segues into a justification for participant observation and how this method can retrieve data to understand hidden financialization processes. *Sentences 2–4* focus on just one ethnographic site. Following the justification for participant observation are the details of the method. Rahman states how he will participate (as an unpaid analyst) and what kind of data that participant observation will retrieve. Notice how the pithy yet extensive description of linked data sets will contribute to answering a specific data-gathering question. *Sentences 5–7* specify more nuanced observation details and their challenges. Here he lays out several aspects of the liquid-making process that will be met with specific field-based interactivities. *Sentence 8* links developer and farmer interactions to specific methodological approaches needed to observe more challenging interactions that may show how market and value become connected.

Sentence 8: I will pay special attention to the different ways that land is valued by these actors and whether and to what effect the details of negotiations circulate in the market.

question focuses just on participant observation. This example shows the breakdown of the methods description, sentence by sentence, simply to give you a sense of how you can narrate the link between your cluster's data sets and the methods. You may choose to replicate this table to do the same, or simply refer to it as you audit your own methods subsections.

Note that in a dissertation or grant proposal, you need to discuss all methods needed to gather data, in order to answer the data-gathering questions. Moreover, make sure that throughout the methods section narrative, you consistently link particular methods with particular data-gathering questions and data sets. This not only justifies their use, but also shows congruence among all project elements.

EXERCISE 10F Maintain Multidimensional Concept Work during Field Research

When you start conducting field research, you will rapidly learn more about your project and the worlds in which it is situated. This means that you will have to adapt its conceptualization throughout the data-collection phase. Knowing in advance that these changes will come is helpful. You can feel at ease knowing that fieldwork requires ongoing concept work and that this strengthens your project's meaningful potential.

Actively staying engaged with your grid in the field allows you to keep track of the relationship between what you thought you were going to do and what you're actually doing. This is a form of ground-truthing. As you stay engaged, you could find that your 4Ps are either not as relevant as they were, or that you have completed your engagement with them and need to stop and take stock of where you are. In doing so, you may sense that there are new people that you need to connect with and that there are new sites and processes to include in your project.

As you write fieldnotes and transcribe interviews, you may find new concepts coming forward while others recede. You may in fact, like both of us have, find that concepts and processes you expected to find in the field don't exist! Ultimately, field research allows you to see how your original project design conceptualization changes into an entirely different project grid that reflects the actual fieldwork itself.

To be consistently engaged with your grid during field research, we suggest that you bring three main concept-work tools to the field with you: The Key Concepts table, concept map, and final research project grid.

1. *Set up the research project grid and other tools.*

    a. Allow your grid, map, and Key Concepts table to be visible by tacking them to a desk, bulletin board, whiteboard, or wall. Know that when your field research begins to flow, you may need to draw on other tools, such as the Data Collecting Interactivities table (9.2), that you used in previous modules.

    b. Create a system that allows you to archive changes in your concept work. You will be absorbing a lot of new information, making the data conceptualization process feel much more quickly paced than prefieldwork design. It may not be until you finish field research and start doing data analysis that some earlier ideas may be important. We've watched students have this experience, coming back to old concept work only to find that it is now more relevant than ever.

2. *Engage daily with the project grid.*

    As you spend your days living among people, engaging with materials that interlocutors have given you, taking photos or recording sounds, writing fieldnotes, and doing other daily creative rituals, check in with your concept-work tools. You can ask, "What continues to be the points of congruence among all the grid and map elements? What is shifting and needs to be newly accounted for?"

Encountering conceptual shifts can be a slow process and make things feel fuzzy. And so it may be difficult to know when to pause and allow insights to surface (which can take weeks or months) or when to act on adjusting and creating congruence among emerging project elements. Whatever the case, you will encounter small and profound conceptual changes. Here's a way to handle shifting elements on a project grid or concept map:

3. *Allow your map or project grid to reveal the new gaps or lack of congruence.* Maybe your data-gathering questions or data sets need to be adjusted. Maybe new aspects of the 4Ps have come into view; maybe your scoping question or significance is shifting. Your concept-work tools will show you how these opportunities for project adjustment can arise, allowing you to return to some of your previous concept work. You can create a running inventory of what needs to be newly addressed or accounted for. You can incorporate these changes by rewriting your RD, double-checking your MO, and reading new literatures. When you make these changes, you can assess whether you need to iterate new maps and project grids. You can also use the first exercise in this module to do a congruence check as the map and grid change over time.

Consistent engagements with concept work like this maintain project congruence and most of all produces a brand-new project grid that will be fundamentally different from your original design work. It will carry you to the next phases of your research project, which are data analysis, as well as dissertation and book writing.

Refining your work, and making new maps and grids for other purposes, will allow you to hold foundational and emerging project intentions, intuitions, dimensions, and intellectual curiosities in a generative resonance. At times this resonance feels almost magical. This is the joyful and enthralling aspect of doing multidimensional work. Multidimensionality becomes a research mode that can be felt and played with, yielding uniquely dynamic works that inform and enliven our worlds.

Resting, Reflecting, Preparing to Begin Anew

You have been on a journey to create a project that constellates multiple personal, creative, and social dimensions. Appreciate how far you have come on this journey: from sketching a research imaginary to completing a project grid. You've crafted your project's framework while supporting colleagues and planning for social life in the field.

We invite you to honor this moment of completion to rest and reflect on some of the skills you cultivated to advance your research goals while staying consciously connected with other aspects of your life and communities. Key to that process was project listening—a mode of pausing that gave you a chance to appreciate and recharge—which can bring ease and joy to the necessary steps of revision and reworking.

One skill we hope you will continue to find useful is the iterative process of connecting your personal curiosity and intuition with your imaginative ethnographic capacities. We centered this handbook around that skill to honor what we have learned about the deeper intentions of people drawn to ethnographic research: that they want to do more than "study people." They often want to imagine new forms and modes of knowledge production. The living project you develop through this process can stay congruent as you, and it, continue to unfold and transform.

We also hope that the handbook's project development process tools—such as research imaginaries and multidimensional objects—can support ethnographic research collectives as they communicate, collaborate, and share their creative works.

Overall, we hope that the activities you have engaged in along the way will help you and your collaborators question the idea that "work" and "life" are separate and oppositional activities.

To this end, this is really just another beginning. We see this handbook as one among many possible creative openings for fostering careful and intentional inquiry-building. One among many ways to conjoin work and life in liberatory ways.

We look forward to learning from you and your work.

APPENDIX 1 Scheduling the Modules for Academic
Quarters and Semesters

10-Week Quarter	Modules
Week 1	Prelude, Introduction, Interlude 1
Week 2	Module 1
Week 3	Module 2
Week 4	Module 3
Week 5	Module 4
Week 6	Module 5
Week 7	Module 6
Week 8	Interlude 2 and Module 7
Week 9	Module 8
Week 10	Module 9
Week 11 (Finals exam)	Read Module 10; do Exercise 10a or revisit previous exercises with guidance from instructor

15-Week Semester	Module
Week 1	Prelude and Introduction
Week 2	Interlude 1
Week 3	Module 1
Week 4	Module 2
Week 5	
Week 6	Module 3
Week 7	Module 4
Week 8	
Week 9	Module 5
Week 10	Module 6
Week 11	Interlude 2 and Module 7
Week 12	
Week 13	Module 8
Week 14	
Week 15	Module 9
Week 16 (Final exams)	Read Module 10; do Exercise 10a or revisit previous exercises with guidance from instructor

Note: We routinely teach all the modules in 10-week quarters, which works very well. If your terms are about fifteen weeks long, then we suggest splitting the exercises for Modules 2, 4, 7, and 8 over two weeks. These modules are the densest and extra time is appropriate for them. The module framework can also support inserting methods training during longer academic terms if that is desirable.

APPENDIX 2 Wilkinson's Partially Filled Research Project Grid

Research Description

The movement against "gender ideology"—a concept used by right-wing groups globally to reference the social construction of gender—is gaining force across Europe and Latin America. Over the past decade, pro-family movements have claimed that gender is rooted in nature and that this scientific fact must be defended. This allows right-wing organizations in Mexico to claim that they must secure the family against gender equality and LGBTQ rights. In doing so, the Mexican pro-family movement makes "gender ideology" a national security problem that bears on state corruption, violence and crime, and neocolonialism. Using ethnographic research, the project follows how right-wing theories of the family connect to theories of the state. It also investigates how gender and security agendas are being shared among authoritarian populist movements across the globe.

Multidimensional Object

Securing the Family

Scoping Question

How does the Mexican pro-family movement's involvement in gender ideology activism, emerging in the context of political and economic security crises, illuminate contemporary forms of transnational right-wing organization and theory?

Significance Statement

This ethnographic project connects ongoing anthropological inquiries into gender, sexuality, nationalism, and security. By doing so, it focuses on the emerging relationship between religious defenses of "traditional" family structures and national security politics. By aiming to interrelate work relevant to Security Studies, Latin American Social Movements, and Gender and Sexuality Studies, it will contribute a timely analysis of how right-wing gender activism figures in the formation of global populist, authoritarian movements tied to national securitization activities.

Key Literatures
Security Studies
Latin American Social Movements
Gender and Sexuality Studies

Process Clusters	Data Sets	Data-Gathering Questions	Specific 4Ps *who/what/where/which processes will you engage to answer this question?*	Methods *how will you get the data?*
Gender	– The genealogy and right-wing theorization of "gender ideology" as a social, political, and anti-Christian construction – Activists' analysis of gender as a security threat, as neocolonial imposition, and as linked to state corruption – State and legislative discourses on equality, women's rights, and LGBT communities	How do pro-family organizations connect their opposition to gender equality and self-determination to processes of moral and scientific nation-building?	– Gray literature produced by pro-family organizations – Pro-family members/activists and organizations; activist meetings, workshops, and protests; activist analysis and strategies; the process of antigender emergence and sustenance – Legislative and state documents on gender and LGBT equity – Allied pro-family institutions activities and gray literature production	– Archival and documentary research on pro-family gray literature and movement materials – Participant observation; oral history interviews; semi-structured interviews – Archival (government) research – Participant observation; archival research

Security	Broad activist discourses and practices that link gender and threats to family and state security How activists personally connect individual, family, and state security How the right-wing theorizes Mexican national crises Mexican state security policies in everyday (education, labor, private property) and national processes (economic, military, migration) Histories of the Mexican state and religious understandings of relationship between family and sovereignty	How do right-wing activists theorize family securitization as a way to combat corruption, create national sovereignty, and build spiritual and public well-being?	Pro-family activists and leadership; the process of connecting gender and security analyses State and media documents and analysis pertaining to security Historical documents, books, and analysis on household and state security Church literature and analysis on family and security	Interviews; participant observation Archival (gray literature) research Archival and historical research Archival and historical research

| The Right-wing | – How people become right-wing activists
– How activists act to connect church and other key institutions and organizations
– The relationship between Mexican pro-family organizations and transnational pro-family organizations—and how they build and exchange information
– Activists' worldviews and political analysis regarding truth, liberalism, science, as well as information-production strategies and logics
– How activists build communication platforms (social media, church networks, etc.) to enable mass mobilization | How are contemporary Mexican right-wing identities, activist programs, and global solidarities being shaped by transnational coordinations to combat gender ideology? | – Pro-family activist members and leadership; their communication strategies, political analysis, and cross-national coalition building
– Pro-family global umbrella organizations | – Surveys; participant observation; life history interviews; archives (gray literature)
– Participant observations at international conferences |

collective agreements Establishing small group agreements regarding how to engage, support, and give feedback. The purpose is to honor and hold space for each other's personal needs and boundaries as well as for the collective dynamic. They also provide guideposts for trustworthy engagement.

collective concept workspace Doing iterative concept work with others in an interpersonally conducive and pro-embodiment space; it helps designers discover project elements and clarify social commitments together. The purpose is to shift research work out of cultures of separation and structural violence and into shared modes of curiosity, connection, and compassion.

concept combos These three to four term combinations bring key ethnographic and theoretical concepts together in a demonstrably creative way, showing that the project offers a new framework of inquiry. Combining concepts in this way contributes to project tensegrity: holding concepts and literatures in productive and innovative tension.

concept map A graphic and nonlinear representation of all project concepts. It displays the project's conceptual connectivity possibilities and, therefore, its emerging multidimensionality.

concept work Making connections between materials and ideas in order to form useful working constructs that lead to generative questions and fieldwork. In project design, concept-work "think tools" help to create and test research feasibility, intention, and potential.

concepts, empirical Terms for specific people, processes, places, things, and ideas that are conversationally and experientially meaningful to your interlocutors and you; concepts that name beings and things that you will directly engage with when you do fieldwork.

concepts, key Specific empirical, broad contextual and theoretical terms that together make up a process. These include the 4Ps and broad contexts.

> **4Ps** Porous groupings of people, parts, processes, and places that need to be directly and ethnographically engaged. The 4Ps help identify and connect with the project's broader contexts.

> **broad contexts** Terms that are often so broad that people either don't directly experience them, or when they do, it is a concept that generalizes a state or experience shared in general (e.g., the state, globalization, global Blackness, security regimes, religious migration). These broad contextual concepts actively shape the spatial and processural flows and frameworks of lived experiences.

concepts, theoretical Generalizable terms and ideas that are important to research communities and the production of knowledge, insight, and liberation; often these are terms to analyze, create, and disseminate new knowledge and representations.

connecting zone See *zones of inquiry*

data-gathering questions See *research questions*

data sets The specific array of information—archival data, interview data, participant observation data, and so on—needed to answer data-gathering questions.

4Ps See *concepts, key*

interacting questions See *research questions*

interacting zone See *zones of inquiry*

literatures Bodies of categorically distinguished work, conventionally bounded by discipline, history, topic, and theme. All literatures have theoretical, topical, methodological, and ideological dimensions that make them meaningful in societies at large. Relating literatures in nonobvious ways is an act of multidimensioning a research project.

multidimensional object (MO) An evocative two- to three-word phrase that connects the research description and the subsequent project elements, including research questions, additional literatures, and other elements. An MO holds the project together through all design stages. It can be expressed with this formula: MO = Ethnographic Concept(s) + Theoretical Concept(s).

multidimensional research design An iterative approach to assembling diverse research concepts and intentions within a congruent framework of inquiry.

multidimensioning The process of defining a project's conceptual combinations and using them to create congruently integrated project elements, from research topic to research questions.

process clusters Are comprised of 4Ps, broad contexts, and other theoretical concepts that can be researched together. Process clusters help you perceive your major datasets.

project congruence Harmonious, intuitive, multidimensional connections established between all project elements such as the MO, RD, research questions, and methods.

project grid A chart that shows multidimensional connections between various parts of a project. The project grid is a flexible frame for concept work and an excellent tool for developing and assessing project balance. It also functions as a shareable snapshot of the project.

project listening The process of intuitively attending to the project's internal elemental coherence and its external relationships with other projects and processes.

research description (RD) A short 100- to 150-word description of what the project is about and its potential significance. The structure of the RD can be expressed with this formula: RD Structure = Research Problem + The Problem's Key Empirical, Contextual, and Theoretical Descriptors + Research Aims

research imaginary A guided, yet freely written narrative about the people, places, parts (objects), processes (4Ps), and scope of a research project. It imagines fieldwork possibilities as well as transformative potentials of the study.

research questions Three different types of research questions are distinguished for their very different forms and functions. All project's questions are directly linked to each other within a multidimensional framework of inquiry. The data collected from interacting questions directly contribute to answering the data-gathering questions, which, in turn, help provide an analytical and theoretical answer to the scoping question.

data-gathering questions Developed out of process clusters and data sets. As empirical questions, they orient work and life in the field; they connect the scoping question to the project's fieldwork plan, broader significance, and otherwise intentions.

interacting questions Specific questions that guide fieldwork involvement in various social processes in order to work with multimodal materials and archives and to converse and engage with people. Field-based interactivities are guided by the data-gathering questions.

scoping question Frames the entire project's ethnographic and theoretical aims. The scoping question can be represented by this formula: Scoping Question = MO + Scoping Terms.

scoping question See *research questions*

scoping zone See *zones of inquiry*

significance, disciplinary Addresses a topical or intellectual gap in the literature and makes an argument as to how the research will address that gap and explain why it is important. Multidimensioning project concepts helps to understand literature gaps in new ways.

significance, social How a research project might contribute to social transformations, such as surfacing the lived experiences of people a researcher works with, catalyzing policy changes, and supporting the emergence of otherwise ways of being and relating.

tensegrity, project Nonhierarchical connectivity that distributes project elements evenly, creating a hanging-togetherness. In contrast to vertical scaling, project tensegrity is a sense-based term to underscore that unusually strong ethnographic projects have parts that may not all seem to belong together in linear or positivistic terms.

vertical scaling Framing sites, things, processes, and people within linear, hierarchical scalar orders and binarized poles, such as "low to high" or "local/global." Such graded ranges can transmit dominant logics of inherent relative value. It can produce reified hegemonic assumptions about intrinsic scalar difference, which invokes images of ranked transcendence. As a result, standard scaling can reify low-high hegemonic notions of difference, development, and value.

zones of inquiry Operationalize the multidimensional object in the form of theoretical, data-collection, and interactive questions. Specifically, they create the conditions for a multidimensional research project when each zone is consistently put in relation to each other.

> **connecting zone** Opens up all the interrelated possibilities (obvious and intuitive) of data collection (directed by process clusters, data sets, and data-gathering questions) in relation to all the project elements so that the scoping question can be connected to the data-gathering questions.
>
> **interacting zone** Focuses on the intellectual and social aspects of fieldwork, allowing the researcher to plan modes of purposeful and ethical inquiry (interacting questions) via observations, interviews, and participation. Interacting questions direct the researcher to gather specific data as well as answer the three to five data-gathering questions.
>
> **scoping zone** Defines the overarching, theoretical, and social field, as well as the project's big (scoping) question and overall significance.

NOTES

Prelude

1   Irani and Silberman, "Stories We Tell about Labor"; Ruha, *Race after Technology*.

2   Vinsel, "*Design Thinking Is a Boondoggle.*"

3   Escobar, *Designs for the Pluriverse*; Suchman, "*Design.*"; Chin,"On Multi-modal Anthropologies from the Space of Design."

4   Penrod and Plastino, *The Dancer Prepares*; Hersey, *Rest as Resistance*; Williams, Owens, and Syedullah, *Radical Dharma*.

5   Cameron, *The Artist's Way*; Chavez, *The Anti-racist Writing Workshop*.

6   Clifford and Marcus, *Writing Culture*; Marcus and Fischer, *Anthropology as Cultural Critique*. Asad, *Anthropology and the Colonial Encounter*; Caulfield, "Culture and Imperialism"; Gough, *Anthropology and Imperialism*; Lewis, "Anthropology and Colonialism"; Said, *Orientalism*; Wagner, *The Invention of Culture*.

7   Abu Lughod, "Writing against Culture"; Harrison, *Decolonizing Anthropology*; Mahmood,*The Politics of Piety*.

8   Boyer, Faubion, and Marcus, *Theory Can Be More Than It Used to Be*; Faubion and Marcus, *Fieldwork Is Not What It Used to Be*; Rabinow, Marcus, Faubion, and Rees, *Designs for an Anthropology of the Contemporary*.

9   Choy, *Ecologies of Comparison*; Behar and Gordon, *Women Writing Culture*; Carr and Lempert, *Scale*; Clarke, *Affective Justice*; Gupta and Ferguson, "Beyond Culture"; Helmreich, *Alien Ocean*; Masco, *The Nuclear Borderlands*; Marcus, *Ethnography through Thick and Thin*; Ong and Collier, *Global Assemblages*; Strathern, *In Relation*; Tsing, *The Mushroom at the End of the World*.

10 Cantarella, Hegel, and Marcus, *Ethnography by Design*; Dumit, *Playing with Methods*; Elliot and Culhane, *A Different Kind of Ethnography*; Hardin and Clarke, *Transforming Ethnographic Knowledge*; Pandian, *A Possible Anthropology*; Sunder Rajan, *Multi-situated*.

11 See for example, Esposito and Evans-Winters, *Introduction to Intersectional Qualitative Research*; Bejarano, Juárez, Garcia, and Goldstein, *Decolonizing Ethnography*; Lincoln, Yvonna S., and Elsa M. Gonzalez y Gonzalez, "The Search for Emerging Decolonizing Methodologies in Qualitative Research"; and Turk, "Suspending Damage."

12 See, among others, brown, *Emergent Strategy*; brown and Imarisha, *Octavia's Brood*; Cox, *Shapeshifters*; Kelly, *Freedom Dreams*; Shange, *Progressive Dystopia*; Sojoyner, "Another Life Is Possible"; and Thomas, *Political Life*.

13 For good guides on pedagogical openness and encouragement, see hooks, *Teaching to Transgress*; and Freire, *Pedagogy of the Oppressed*.

14 Weston, *Families We Choose*.

Introduction: Multidimensional Concept Work

1 By *congruent* we mean diverse project concepts that together express a sense of agreement or harmony; this is opposed to *coherence*, which conveys a more rigid quality of the Western logics of consistency.

2 For different perspectives of anthropology and the otherwise, including abolition otherwise, see, among others, Berry, Argüelles, Cordis, Ihmoud, and Estrada, "Toward a Fugitive Anthropology"; McTighe and Raschig, "An Otherwise Anthropology"; Meek and Morales Fontanilla, "Otherwise"; Povinelli, "Routes/Worlds"; and Restrapo and Escobar, "Other Anthropologies and Anthropology Otherwise."

3 Sanabria, "Imagining Otherwise Encounters after Epistemicide," 5; emphasis in original.

4 Seminal works that support this attitude include writings by, among others, Jean-Paul Sartre, Jacques Lacan, and Charles Taylor.

5 Akómoláfé, *The Allegory of the Pit*; Chandler, *Toward an African Future*; Ferreira da Silva, *A Global Idea of Race*; Wynter, "Unsettling the Coloniality of Being/Power/Truth/Freedom."

6 We are cautioned here by Báyò Akómoláfé's analysis that even when one tries to uninhabit these liberal frameworks, researchers can reproduce ontoepistemologies of Man (Wynter, "Unsettling the Coloniality of Being/Power/Truth/Freedom")—the very thing often being, ironically, refuted (Akómoláfé, *The Allegory of the Pit*).

7 Glissant, *Poetics of Relation*. We are referring here to European imperial languages like English, French, or German.

8 Hurston, *Dust Tracks on a Road*; Hurston, *Of Mules and Men*.

9   Mills, *The Sociological Imagination*.

10   See, among others, *The Asthma Files*; Boellstorff, Nardi, Pearce, and Taylor, *Ethnography and Virtual Worlds*; Cantarella, Hegel, and Marcus, *Ethnography by Design*; Chin, "On Multimodal Anthropologies from the Space of Design"; Dumit, "Writing the Implosion"; Elliott and Culhane, *A Different Kind of Ethnography*; and Holmes and Marcus, "Para-ethnography and the Rise of the Symbolic Analyst."

11   Ballestero and Winthereik, *Experimenting with Ethnography*; Centre for Imaginative Ethnography, https://imaginative-ethnography.com; Murphy and Marcus, "Epilogue"; Nye and Hamdy, "Drawing the Revolution."

12   For works regarding marronage and fugitivity, see Gordon, *The Hawthorne Archive*; Harney and Moten, *The Undercommons*; Harrison, *Decolonizing Anthropology*; and Robinson, *Black Movements in America*.

For feminist approaches to a range of anthropological inquiry topics, see Craven and Davis, *Feminist Activist Ethnography*; Davis and Craven, *Feminist Ethnography*; Fleuhr-Lobban, "Collaborative Anthropology"; Lewin and Silverstein, *Mapping Feminist Anthropology in the Twenty-First Century*; and McClaurin, *Black Feminist Anthropology*.

For studies of Indigenous political cultures, see Coulthard, "For Our Nations to Live, Capitalism Must Die"; Wildcat, McDonald, Irlbacher-Fox, and Coulthard, "Learning from the Land"; A. Simpson, *Mohawk Interruptus*; L. Simpson, *As We Have Always Done*; and Tuck and Yang, "Decolonization Is Not a Metaphor."

For multispecies work, see Akómoláfé, *I, Coronavirus*; Gagliano, *Thus Spoke the Plant*; Gumbs, *Undrowned*; Haraway, *When Species Meet*; Kimmerer, *Braiding Sweetgrass*; Kohn, *How Forests Think*; and Parreñas, *Decolonizing Extinction*.

13   Dara Culhane, in the introduction to *A Different Kind of Ethnography*, asserts that, "when approached as a process or practice, as something relational and productive, imagination leads to new spaces of inquiry, spaces that are dependent upon the collaborative nature of anthropological knowledge. Such an approach situates imagination as a pedagogy, and one with the potential to open up and to make visible the unknown," 16.

14   Morris, "Where It Hurts," 540.

15   Fanon, *Black Skin, White Masks*, 229.

16   Batchelor, *Faith and the Imagination in Dharma Practice*.

17   brown and Imarisha, *Octavia's Brood*, 4.

18   For a review of anthropology and the imagination, see Sneath, Holbraad, and Pedersen, "Technologies of the Imagination."

19   Hobart, "At Home on the Mauna"; Strathern, *The Relation*; Todd, "Fish Plu-ralities," 217; and Tsing, *The Mushroom at the End of the World*. See also Tsing, "On Nonscalability," and Trouillot, *Silencing the Past*.

20   See Bhabha, *The Location of Culture*, and Harrison, "Theorizing in Ex-centric Sites."

21   We thank Emilia Sanabria for this helpful metaphor!

22   We especially follow Michel-Rolph Trouillot's (2001) critique of "object of study" and what we view as his multidimensional reconceptionalization of this term.

23   Thanks to Joe Dumit for helping us think with tensegrity.

24   Pugh, *An Introduction to Tensegrity*; Snelson, "The Art of Tensegrity."

25   For inspiration on staying present in community for liberatory times, see Prentis Hemphill's podcast, *Finding Our Way*, and the website of the Embodiment Institute, theembodimentinstitute.org.

Interlude 1: Creating a Collective Concept Workspace

1   adrienne marie brown, interview with Prentis Hemphill. The Emergent Strategy podcast, May 27, 2021, https://open.spotify.com/episode/3BkJglLio5svyluQbmxqck?si=m9wuoqXxT3GL_PsPo4Zjug&nd=1.

2   adrienne marie brown, interview with Prentis Hemphill. The Emergent Strategy podcast, May 27, 2021, https://open.spotify.com/episode/3BkJglLio5svyluQbmxqck?si=m9wuoqXxT3GL_PsPo4Zjug&nd=1

3   For further ideas on agreements, see the East Bay Meditation Center Agreements, https://eastbaymeditation.org/2022/03/agreements-for-multicultural-interactions/. Accessed August 10, 2023. Note that practicing these agreements helps to hone participant observation and interview skills as well.

4   Writers at Work, https://writersatwork.com.

5   Chavez, *The Anti-racist Writing Workshop*; Lerman, *The Critical Response Process*.

Module 1: Imagine the Research

1   Behar, *The Vulnerable Observer*; Hurston, *Dust Tracks on a Road*; Powdermaker, *Stranger and Friend*; Strathern, *The Gender of the Gift*.

Module 2: Focus on Literatures

1   Ngũgĩ wa Thiong'o, *Decolonising the Mind*.

Module 3: Map Concepts

1   Other humanities and social science scholars deploy mapping, usually in the analysis phases; for a critical theoretical approach to mapping contexts for analysis, see Clarke, Friese, and Washburn, *Situational Analysis*. Before creating our own concept maps, we learned a lot from Maxwell, *Qualitative Research Design*, 54–63.

Module 6: Perceive Your Multidimensional Object

1   Fortun, *Advocacy after Bhopal*; Ong, *Flexible Citizenship*; Parreñas, *Decolonizing Extinction*; Shange, *Progressive Dystopia*.

2   Olson, *Into the Extreme*; Peterson, *Speculative Markets*.

Module 7: The Scoping Zone

1   Dumit, "Writing the Implosion."

BIBLIOGRAPHY

Abu Lughod, Lila. "Writing against Culture." In *Recapturing Anthropology: Working in the Present*, edited by Richard G. Fox, 137–162. Santa Fe, NM: School of American Research Press, 1991.

Akómoláfé, Báyò. *The Allegory of the Pit: Or the Irony of Victory*. Accessed June 9, 2023. https://www.bayoakomolafe.net/post/the-allegory-of-the-pit-or-the-irony-of-victory.

Akómoláfé, Báyò. *I, Coronavirus, Mother, Monster, Activist*. Accessed June 9, 2023. https://www.bayoakomolafe.net/post/i-coronavirus-mother-monster-activist.

Asad, Talal. *Anthropology and the Colonial Encounter*. London: Ithaca Press, 1973.

The Asthma Files. Accessed June 13, 2021. https://theasthmafiles.org.

Ballestero, Andrea, and Brit Ross Winthereik, eds. *Experimenting with Ethnography: A Companion to Analysis*. Durham, NC: Duke University Press, 2021.

Batchelor, Stephen. *Faith and the Imagination in Dharma Practice: A Seminar with Stephen Batchelor*. Barre, MA: Barre Center for Buddhist Studies, January 23, 2021.

Behar, Ruth. *The Vulnerable Observer: Anthropology That Breaks Your Heart*. Boston: Beacon, 1996.

Behar, Ruth, and Deborah A. Gordon. *Women Writing Culture*. Berkeley: University of California Press, 1996.

Bejarano, Carolina Alonso, Lucia López Juárez, Mirian A. Mijangos Garcia, and Daniel M. Goldstein. *Decolonizing Ethnography: Undocumented Immigrants and New Directions in Social Science Research*. Durham, NC: Duke University Press, 2019.

Benjamin, Ruha. *Race after Technology: Abolitionist Tools for the New Jim Code*. Oxford: Polity, 2019.

Berry, Maya J., Claudia Chávez Argüelles, Shanya Cordis, Sarah Ihmoud, and Elizabeth Velásquez Estrada. "Toward a Fugitive Anthropology: Gender, Race, and Violence in the Field." *Cultural Anthropology* 32, no. 4 (2017): 537–65.

Bhabha, Homi. *The Location of Culture*. Milton Park, UK: Routledge, 2004.

Boellstorff, Tom, Bonnie Nardi, Celia Pearce, and Tina L. Taylor. *Ethnography and Virtual Worlds: A Handbook of Method.* Princeton, NJ: Princeton University Press, 2012.

Boyer, Dominic, James D. Faubion, and George E. Marcus. *Theory Can Be More Than It Used to Be: Learning Anthropology's Method in a Time of Transition.* Ithaca, NY: Cornell University Press, 2015.

brown, adrienne marie. *Emergent Strategy: Shaping Change, Changing Worlds.* Chico, CA: AK Press, 2017.

brown, adrienne marie, and Walidah Imarisha, eds. *Octavia's Brood: Science Fiction Stories from Social Justice Movements.* Chico, CA: AK Press, 2015.

brown, adrienne marie. Interview with Prentis Hemphill. The Emergent Strategy podcast. May 27, 2021. https://open.spotify.com/episode /3BkJglLio5svyIuQbmxqck?si=m9wuoqXxT3GL_PsP04Zjug&nd=1.

Cameron, Julia. *The Artist's Way: A Spiritual Path to Higher Creativity.* New York: TarcherPerigee, 2016.

Cantarella, Luke, Christine Hegel, and George E. Marcus. *Ethnography by Design: Scenographic Experiments in Fieldwork.* Milton Park, UK: Routledge, 2019.

Carr, E. Summerson, and Michael Lempert. *Scale: Discourse and Dimensions of Social Life.* Berkeley: University of California Press, 2016.

Caulfield, Mina D. "Culture and Imperialism: Proposing a New Dialectic." In *Reinventing Anthropology*, edited by Dell Hymes, 182–212. New York: Vintage, 1974.

Chandler, Nahum. *Toward an African Future: Of the Limit of World.* Albany: State University of New York Press, 2021.

Chavez, Felicia Rose. *The Anti-racist Writing Workshop: How to Decolonize the Creative Classroom.* Chicago: Haymarket Books, 2021.

Chin, Elizabeth. "On Multimodal Anthropologies from the Space of Design: Toward Participant Making." *American Anthropologist* 119 (2017): 541–43.

Choy, Tim. *Ecologies of Comparison: An Ethnography of Endangerment in Hong Kong.* Durham, NC: Duke University Press, 2011.

Clarke, Adele E., Carrie Friese, and Rachel S. Washburn. *Situational Analysis: Grounded Theory after the Postmodern Turn.* Thousand Oaks, CA: Sage, 2017.

Clarke, Kamari. *Affective Justice: The International Criminal Court and the Pan-Africanist Pushback.* Durham, NC: Duke University Press, 2019.

Clifford, James, and George E. Marcus, eds. *Writing Culture: The Poetics and Politics of Ethnography.* Berkeley: University of California Press, 1986.

Coulthard, Glen. "For Our Nations to Live, Capitalism Must Die." *Unsettling America: Decolonization in Theory and Practice.* November 5, 2013. https:// unsettlingamerica.wordpress.com/2013/11/05/for-our-nations-to-live -capitalism-must-die/.

Cox, Aimee. *Shapeshifters: Black Girls and the Choreography of Citizenship.* Durham, NC: Duke University Press, 2015.

Craven, Crista, and Dána-Ain Davis. *Feminist Activist Ethnography: Counterpoints to Neoliberalism in North America*. Lanham, MD: Lexington Books, 2013.

Culhane, Dara. "Introduction." In *A Different Kind of Ethnography: Imaginative Practices and Creative Methodologies*. Toronto: University of Toronto Press, 2016.

Davis, Dána-Ain, and Crista Craven, eds. *Feminist Ethnography: Thinking through Methodologies, Challenges and Possibilities*. Lanham, MD: Rowman and Littlefield, 2022.

Dumit, Joseph. *Playing with Methods: Messing with Thinking, Writing, and Learning Together*. Durham, NC: Duke University Press, forthcoming.

Dumit, Joseph. "Writing the Implosion: Teaching the World One Thing at a Time." *Cultural Anthropology* 29, no. 2 (2014): 344–362. https://journal.culanth.org /index.php/ca/article/view/ca29.2.09/301.

Elliott, Denielle, and Dara Culhane. *A Different Kind of Ethnography: Imaginative Practices and Creative Methodologies*. Toronto: University of Toronto Press, 2016.

Embodiment Institute. theembodimentinstitute.org.

Escobar, Arturo. *Designs for the Pluriverse: Radical Interdependence, Autonomy, and the Making of Worlds*. Durham, NC: Duke University Press, 2018.

Esposito, Jennifer and Venus Evans-Winters. *Introduction to Intersectional Qualitative Research*. Thousand Oaks, CA: Sage Publications, 2021.

Fanon, Frantz. *Black Skin, White Masks*. New York: Grove Press, 2008.

Faubion, James D., and George E. Marcus, eds. *Fieldwork Is Not What It Used to Be: Learning Anthropology's Method in a Time of Transition*. Ithaca, NY: Cornell University Press, 2009.

Ferreira da Silva, Denise. *A Global Idea of Race*. Minneapolis: University of Minnesota Press, 2008.

Fleuhr-Lobban, Carolyn. "Collaborative Anthropology as 21st Century Ethical Anthropology." *Collaborative Anthropologies* 1 (2008): 176–82.

Fortun, Kim. *Advocacy after Bhopal: Environmentalism, Disaster, New Global Orders*. Chicago: University of Chicago Press, 2001.

Freire, Paulo. *Pedagogy of the Oppressed*. New York: Continuum, 2000.

Gagliano, Monica. *Thus Spoke the Plant: A Remarkable Journey of Groundbreaking Scientific Discoveries and Personal Encounters with Plants*. Berkeley, CA: North Atlantic Books, 2018.

Glissant, Édouard. *Poetics of Relation*. Ann Arbor: University of Michigan Press, 1997.

Gordon, Avery. *The Hawthorne Archive: Letters from the Utopian Margins*. New York: Fordham University Press, 2017.

Gough, Kathleen. *Anthropology and Imperialism*. Radical Education Project, 1967.

Gumbs, Alexis Pauline. *Undrowned: Black Feminist Lessons from Marine Mammals*. Oakland, CA: AK Press, 2020.

Gupta, Akhil, and James Ferguson. "Beyond 'Culture': Space, Identity, and the Politics of Difference." *Cultural Anthropology* 7, no. 1 (1992): 6–23.

Haraway, Donna. *When Species Meet*. Minneapolis: University of Minnesota Press, 2007.

Hardin, Rebecca, and Kamari Maxine Clarke, eds. *Transforming Ethnographic Knowledge*. Madison: University of Wisconsin Press, 2012.

Harney, Stefano, and Fred Moten. *The Undercommons: Fugitivity and Black Study*. Wivenhoe / New York / Port Watson: Minor Compositions, 2013.

Harrison, Faye V. *Decolonizing Anthropology: Moving Further toward an Anthropology for Liberation*. Arlington, VA: American Anthropological Association, 2011.

Harrison, Faye V. "Theorizing in Ex-centric Sites." *Anthropological Theory* 16, nos. 2–3 (2016): 160–76.

Helmreich, Stefan. *Alien Ocean: Anthropological Voyages in Microbial Seas*. Berkeley: University of California Press, 2009.

Hemphill, Prentis. *Finding Our Way* podcast. https://www.findingourwaypodcast.com.

Hersey, Tricia. *Rest as Resistance*. New York: Little, Brown Spark, 2022.

Hobart, Hi'ilei Julia. "At Home on the Mauna: Ecological Violence and Fantasies of Terra Nullius on Maunakea's Summit." *Native American and Indigenous Studies* 6, no. 2 (2019): 30–50.

Holmes, Douglas R., and George E. Marcus. "Para-ethnography and the Rise of the Symbolic Analyst." In *Frontiers of Capital: Ethnographic Reflections on the New Economy*, edited by Melissa S. Fisher and Greg Downey, 33–57. Durham, NC: Duke University Press, 2006.

hooks, bell. *Teaching to Transgress: Education as the Practice of Freedom*. Abingdon, UK: Routledge, 1994.

Hurston, Zora Neale. *Dust Tracks on a Road*. New York: HarperPerennial, [1942] 1991.

Hurston, Zora Neale. *Of Mules and Men*. New York: Perennial Library, 1935.

Irani, Lilly, and M. Six Silberman. "Stories We Tell about Labor: Turkopticon and the Trouble with 'Design.'" San Diego: UC San Diego, 2016. https://escholarship.org/uc/item/8nm273g3.

Kelly, Robin D.G. *Freedom Dreams: The Black Radical Imagination*. Boston: Beacon, 2003.

Kimmerer, Robin Wall. *Braiding Sweetgrass: Indigenous Wisdom, Scientific Knowledge and the Teachings of Plants*. Minneapolis, MN: Milkweed Editions, 2015.

Kohn, Eduardo. *How Forests Think: Toward an Anthropology beyond the Human*. Berkeley: University of California Press, 2013.

Lerman, Liz. *The Critical Response Process*. https://lizlerman.com/critical-response-process.

Lewin, Ellen, and Leni M. Silverstein, eds. *Mapping Feminist Anthropology in the Twenty-First Century*. Brunswick, NJ: Rutgers University Press, 2016.

Lewis, Diane. "Anthropology and Colonialism." *Current Anthropology* 14, no 1 (1973): 581–602.

Lincoln, Yvonna S., and Elsa M. Gonzalez y Gonzalez. "The Search for Emerging Decolonizing Methodologies in Qualitative Research: Further Strategies for Liberatory and Democratic Inquiry." *Qualitative Inquiry* 14, no. 5 (2008): 784–805.

Mahmood, Saba. *The Politics of Piety: The Islamic Revival and the Feminist Subject.* Princeton, NJ: Princeton University Press, 2011.

Marcus, George E. *Ethnography through Thick and Thin.* Princeton, NJ: Princeton University Press, 1998.

Marcus, George E., and Michael Fischer. *Anthropology as Cultural Critique: An Experimental Moment in the Human Sciences.* Chicago: University of Chicago Press, 1986.

Masco, Joseph. *The Nuclear Borderlands: The Manhattan Project in Post–Cold War New Mexico.* Princeton, NJ: Princeton University Press, 1999.

Maxwell, Joseph A. *Qualitative Research Design: An Interactive Approach.* Thousand Oaks, CA: Sage, 2012.

McClaurin, Irma. *Black Feminist Anthropology: Theory, Politics, Praxis, and Poetics.* Black Women Writers Series. New Brunswick, NJ: Rutgers University Press, 2001.

McTighe, Laura, and Megan Racshid. "An Otherwise Anthropology." *Cultural Anthropology*, Editor's Forum: Theorizing the Contemporary. July 31, 2019. https://culanth.org/fieldsights/series/an-otherwise-anthropology.

Meek, Laura A., and Julia Alejandra Morales Fontanilla. "Otherwise." *Feminist Anthropology* 3, no. 2 (2022): 274–83. https://doi.org/10.1002/fea2.12094.

Mills, C. Wright. *The Sociological Imagination.* Oxford: Oxford University Press, 2000.

Morris, Courtney Desiree. "Where It Hurts: 2014 Year in Review." *American Anthropologist* 117, no. 3 (2015): 540–52.

Murphy, Keith, and George E. Marcus. "Epilogue: Ethnography and Design, Ethnography in Design . . . Ethnography by Design." In *Design Anthropology: Theory and Practice*, edited by Wendy Gunn, Ton Otto, and Rachel Charlotte Smith, 251–68. London: Bloomsbury Academic, 2013.

Ngũgĩ wa Thiong'o. *Decolonising the Mind: The Politics of Language in African Culture.* Portsmouth, NH: Heinemann Educational Publishing, 1986.

Nye, Coleman, and Sherine Hamdy. "Drawing the Revolution: The Practice and Politics of Collaboration in the Graphic Novel *Lissa*." *Ada: A Journal of Gender, New Media, and Technology* 14 (2014). https://doi.org/10.5399/uo/ada.2018.14.5.

Olson, Valerie. *Into the Extreme: US Environmentalism and Politics beyond Earth.* Minneapolis: University of Minnesota Press, 2018.

Ong, Aihwa. *Flexible Citizenship: The Cultural Logics of Transnationality*. Durham, NC: Duke University Press, 1999.

Ong, Aihwa, and Stephen Collier, eds. *Global Assemblages: Technology, Politics and Affect*. Hoboken, NJ: Wiley-Blackwell, 2005.

Pandian, Anand. *A Possible Anthropology: Methods for Uneasy Times*. Durham, NC: Duke University Press, 2019.

Parreñas, Juno Salazar. *Decolonizing Extinction: The Work of Care in Orangutan Rehabilitation*. Durham, NC: Duke University Press, 2018.

Penrod, James, and Janice Gudde Plastino. *The Dancer Prepares: Modern Dance for Beginners*. New York: McGraw-Hill Education, 2004.

Peterson, Kristin. *Speculative Markets: Drug Circuits and Derivative Life in Nigeria*. Durham, NC: Duke University Press, 2014.

Povinelli, Elizabeth A. "Routes/Worlds." *e-flux*, no. 27 (2011).

Powdermaker, Hortense. *Stranger and Friend: The Way of an Anthropologist*. New York: W. W. Norton, 1966.

Pugh, Anthony. *An Introduction to Tensegrity*. Berkeley: University of California Press, 1976.

Rabinow, Paul, George E. Marcus, James D. Faubion, and Tobias Rees. *Designs for an Anthropology of the Contemporary*. Durham, NC: Duke University Press, 2008.

Restrapo, Eduardo, and Arturo Escobar. "Other Anthropologies and Anthropology Otherwise." *Critique of Anthropology* 25, no. 2 (2005): 99–129.

Robinson, Cedric. *Black Movements in America*. Milton Park, UK: Routledge, 1997.

Said, Edward. *Orientalism*. New York: Vintage, 1979.

Sanabria, Emilia. "Imagining Otherwise Encounters after Epistemicide." Presented at Plantas Sagradas en las Americas II (Sacred Plants in the Americas II) Chacruna Global Psychedelic Summit, April 25, 2021.

Shange, Savannah. *Progressive Dystopia: Abolition, Antiblackness, and Schooling in San Francisco*. Durham, NC: Duke University Press, 2019.

Simpson, Audra. *Mohawk Interruptus: Political Life across the Borders of Settler States*. Durham, NC: Duke University Press, 2014.

Simpson, Leanne Betasamosake. *As We Have Always Done: Indigenous Freedom through Radical Resistance*. Minneapolis: University of Minnesota Press, 2017.

Sneath, David, Martin Holbraad, and Morten Axel Pedersen. "Technologies of the Imagination: An Introduction." *Ethnos* 74, no. 1 (2009): 5–30.

Snelson, Kenneth. "The Art of Tensegrity." *International Journal of Space Structures* 27, nos. 2–3 (2012): 71–80.

Sojoyner, Damien M. "Another Life Is Possible: Black Fugitivity and Enclosed Places." *Cultural Anthropology* 32, no. 4 (2017): 514–53.

Strathern, Marilyn. *The Gender of the Gift: Problems with Women and Problems with Society in Melanesia*. Berkeley: University of California Press, 1988.

Strathern, Marilyn. *In Relation: An Anthropological Account*. Durham, NC: Duke University Press, 2020.

Strathern, Marilyn. *The Relation: Issues in Complexity and Scale*. Cambridge: Cambridge University Press, 1995.

Suchman, Lucy. "Design." Theorizing the Contemporary, *Fieldsights*, March 29, 2018. https://culanth.org/fieldsights/design.

Sunder Rajan, Kaushik. *Multi-situated: Ethnography as Diasporic Praxis*. Durham, NC: Duke University Press, 2021.

Thomas, Deborah. *Political Life in the Wake of the Plantation*. Durham, NC: Duke University Press, 2019.

Todd, Zoe. "Fish Pluralities: Human-Animal Relations and Sites of Engagement in Paulatuuq, Arctic Canada." *Études/inuit/studies* 38, no. 1–2 (2014): 217–38.

Trouillot, Michel-Rolph. "The Anthropology of the State in the Age of Globalization: Close Encounters of the Deceptive Kind." *Current Anthropology* 42, no.1 (2001): 125–38.

Trouillot, Michel-Rolph. *Silencing the Past: Power and the Production of History*. Boston: Beacon, 2015.

Tsing, Anna Lowenhaupt. *The Mushroom at the End of the World: On the Possibility of Life in Capitalist Ruins*. Princeton, NJ: Princeton University Press, 2015.

Tsing, Anna Lowenhaupt. "On Nonscalability: The Living World Is Not Amenable to Precision-Nested Scales." *Common Knowledge* 18, no. 3 (2012): 505–24.

Tuck, Eve. "Suspending Damage: A Letter to Communities." *Harvard Educational Review* 79, No. 3 (2009): 409–27.

Tuck, Eve, and K. Wayne Yang. "Decolonization Is Not a Metaphor." *Tabula Rasa* 38 (2021): 61–111.

Vinsel, Lee. "Design Thinking Is a Boondoggle." *Chronicle of Higher Education*, May 21 2018. https://www.chronicle.com/article/Design-Thinking-Is-a/243472, 2018.

Wagner, Roy. *The Invention of Culture*. Chicago: University of Chicago Press, 1975.

Weston, Kath. *Families We Choose: Lesbians, Gays, Kinship*. New York: Columbia University Press, 1997.

Wildcat, Matthew, Mandee McDonald, Stephanie Irlbacher-Fox, and Glen Coulthard. "Learning from the Land: Indigenous Land-Based Pedagogy and Decolonization." *Decolonization: Indigeneity, Education and Society* 3, no. 3 (2014): i–xv.

Williams, Rev. angel Kyodo, Lama Rod Owens, with Jasmine Syedullah. *Radical Dharma: Talking Race, Love, and Liberation*. Berkeley, CA: North Atlantic Books, 2016.

Wynter, Sylvia. "Unsettling the Coloniality of Being/Power/Truth/Freedom: Towards the Human, after Man, Its Overrepresentation—an Argument." cr: *The New Centennial Review* 3, no. 3 (2003): 257–337.

INDEX

Academia (search engine), 128

active inquiry words, 210

agreements: collective, 37–42, 309; collective concept workspace and, 35–36; community, 37–39

Akómoláfé, Báyò, 9

Annual Review of Anthropology (journal), 83

anonymity expectations, 276

AnthroSource, 127

The Anti-racist Writing Workshop (Chavez), xx

archival questions, 248–49

archive: accounting for, 261–64; data sets-data-gathering questions-archives relationship, 263, 266; interacting questions and, 247, 250, 254; preparing for, 97–101; questions about, 267; reflections in, 55; Research Project Grid and, 299; workspace preparation for, 33–34

area studies, 75

The Artist's Way (Cameron), xx

assumptions, 55–57

autobiography, 44

Badami, Nandita, 151–52, 176, 179

balanced combos, 124–26

Ballestero, Andrea, 10

Behar, Ruth, 44

biographical questions, 267

biotensegrity, 20

Boellstorff, Tom, 74

boundaries, 35

brainstorming: concept maps and, 97; literatures diagram and, 89; multidimensional object (MO) and, 180, 182

broad contexts, 52, 66, 124, 310; assumptions and, 56; disconnected, 58; expanding and revising, 87; identifying, 54; in concept map, 101–2; key concepts table and, 62, 118, 120; multidimensional object (MO) and, 178–79, 184–85; in scoping tables, 200

brown, adrienne marie, 35

bubble maps, 98

Cameron, Julia, xx

case studies, 217

Center for Imaginative Ethnography, 10

Chavez, Felicia Rose, xx, 40

check-ins, 34–35

citations: concept combos and, 128–30; congruence in, 132; finding key literatures and, 79; future reading lists and, 92; multidimensional object (MO) or research description (RD) and, 292–93; sharing, 91; through-putting, 292

citizenship documents, 117

coherence, 316n1

collective agreements, 309; making, 37–42

collective concept work, 31–32

collective concept workspace, 31–32, 309; agreements and, 35–36; closing or continuing, 271–72; concept combos and, 138; concept map and, 109–10; interacting zone and, 270–71; key literatures and, 93–94; multidimensional object (MO)

connecting zone and, 267–68; conversational interactivity questions, 257–61; establishing data collecting interactivities, 250–55; preparing for archive and, 262–64; three *hows* and, 248–49, 254

interdependence, 32

internal spaciousness, 25–26

interpersonal interactivities, 248

interrelational tension, 20

interviews, 252–53; assumptions and, 57–58; conversational interactivities and, 257–61; guiding questions for, 258–59; interacting questions and, 260–61; nonstandard and other methods and, 264–65; research descriptions (RDS) and, 162

intuition, 6–8, 10–11, 55, 58

iterative process, 3, 8, 112, 301; in multidimensional design process, 11; valuing, xxiv–xxvi

key concepts, 2–3, 44, 310; contracting, 152; creating table of, 62–64; cross-cutting combo opportunities, 126; expanding by including literature based others, 86–89; field research and, 299; finding key literatures and, 76–77, 81; limiting number of, 88; literature categories and, 71, 80; multidimensional object (MO) and, 178–79; placing within literatures, 116–23; research description (RD) structure and, 147–48; research imaginary and, 51–54; revising, 170; revising table and map of, 136; scoping questions and, 198; in scoping tables, 200; search terms from, 77; Venn diagrams for, 114, 117, 121, 123, 126, 153

key descriptors, 147–48, 174; concept combos, 157; identifying, 152–57; interrelating, 157–61; multidimensionality and, 157–61

key literatures, 132; collective concept workspace and, 93–94; finding, 76–83; literature review and, 286, 290; project in conversations with, 83–86; Research Project Grid and, 92; updating, 133

Kladky, Ellen, 150–51, 175–76; conversational interactivity questions, 258–59; data-gathering questions, 240–42; data sets, 234; process clusters, 228, 234–36

knowledge siloing, 70

legal anthropology, 80

Lerman, Liz, 40

library catalogs, 84

lifeworlds, 45

linear scalar schemas, 12

listening: concept maps and, 96, 108–9; to feedback, 41; in five "whys," 219; to literatures diagram, 89–90; to literature work, 76; mindful, 39; pausing for creativity and, 170–72; project, 25–26, 54, 83, 96, 141–43, 311; reflections and, 54; slow reading as, 60; tensegrity and, 26

literature-based other concepts, 86–89

literature reviews, 128; concept combos from, 157; in project proposal, 286–94

literatures, 69, 310; area studies and, 75; category identification, 78–79; concept combos and, 124, 129–30; concepts and, 72–76; defining, 72; diagram for, 89–90; expected and unexpected, 73–74; finding with project concepts, 76–83; gaps in, 286, 291–92; key concept placement within, 116–23; normative categories of, 70; preliminary reading list, 90–92; project concepts and, 70, 76–83; project in conversations with, 83–86; research description (RD) and, 155; strategic review of, 83–86; substitution or addition decisions, 131–32; Venn diagram for, 74–75, 112, 116, 122–23, 131–32, 291; whole-project congruence check and, 275 literature searches: concept combos and, 127–34; key concepts and terms for, 77

local/global scale, 13

main overarching question, 196–97

Mallin, Sean, 279, 287, 289, 292

Marcus, George, 10, 43

material dimensions, 3

materiality, 131

McKinson, Kimberley, 23, 117, 124, 158; concept-combo literature work, 132; concept combos, 125, 134; concept placements, 119; data sets-data-gathering questions-archives relationship, 263; key concepts, 118; scoping question, 206–8

McLaughlin-Alcock, Colin, 265–66

meanings: conceptual categories and, 104; feedback and, 40

medical anthropology, 89–90, 132, 290

Medical Anthropology Quarterly, 132

memoirs, 44

methods: data sets-data-gathering questions-methods relationship, 251–54, 256; emergent, 264; Ethics Review Board application and, 276; nonstandard, 264–66; other, 264–67; project, 249; project proposal section for, 294–98; standard, 264

methods column, 251

Mills, C. Wright, 10

mindful listening, 39

mindful speaking, 39

mind-mapping programs, 100

MO. *See* multidimensional object

Morris, Courtney, 11

motivation: personal, 219; research imaginary and, 56; scoping question and, 210

multidimensional concept work, 6; during field research, 298–300

multidimensional design, xviii–xx, xxiii, 3, 5; community necessity, xxvi; defining, 2; elements and processes of, 26–29; imaginative and iterative process in, 11; valuing iterative process in, xxiv–xxvi

multidimensionality, 4, 18; concept combos and, 113; engaging, in research imaginary, 54–59; imagining, 59; iteratively building, 112; key descriptors and, 157–61; research description (RD) and, 157; in reflecting on review articles, 85

multidimensional object (MO), 116, 167, 203, 310; brainstorming, 180, 182; collective concept workspace, 189–90; comparing candidates for, 183; congruence and, 184–85; connecting zone and, 224; defining, 19; feeling into, 188; formula for, 173; framework of inquiry and, 168, 184–85; getting unstuck in, 187–88; identifying, 19–24; literature review and, 286, 292–93; materials and space for, 180–83; pausing for creativity in, 170–72; perceiving and creating, 180–88; possibilities of, 185; process clusters and, 231; proposal opening paragraph and, 279, 281–83, 285; research description (RD) relationship to, 173–80; reflecting and selecting, 184–88; Research Project Grid and, 188; scoping question and, 206; scoping tables and, 199–200; whole-project congruence check and, 275

multidimensional research design, 311; customized methods and, 264

multidimensional scaling: conceptual frameworks and, 15; perceptual space and, 15; unhoused individuals and, 16–17

multidimensional space, getting into, 12–19

multidimensional tensegrity, 173–80

multidimensional zones of inquiry, 24–25

multidimensioning, 2, 5, 112, 158, 223, 311; concept combo creation, 123–27

Murphy, Keith, 10

narration, 9

narratives: concept map explanation with, 109; key concepts table and,

Printed and bound by CPI Group (UK) Ltd, Croydon, CR0 4YY

16/04/2025

14658730-0001